CREATURE COMFORT

Animals That Heal

BERNIE GRAHAM

Medical Consultant: Hyman Gross, M.D.
Clinical Associate Professor of Neurology at the
University of Southern California

 **Prometheus Books**

59 John Glenn Drive
Amherst, New York 14228-2197

First published in Great Britain by Simon & Schuster UK Ltd., 1999
A Viacom company

Simon & Schuster UK Ltd
Africa House
64–78 Kingsway
London WC2B 6AH
United Kingdom

Copyright United Kingdom © Bernie Graham, 1999,
under the Berne Convention

Published 2000 by Prometheus Books

Inquiries should be addressed to
Prometheus Books, 59 John Glenn Drive, Amherst, New York 14228–2197.
VOICE: 716–691–0133, ext. 207.
FAX: 716–564–2711.
WWW.PROMETHEUSBOOKS.COM

04 03 02 01 00   5 4 3 2 1

Library of Congress Cataloging-in-Publication Data

Graham, Bernie, 1959–
    Creature comfort : animals that heal / Bernie Graham.
        p.   cm.
    Includes bibliographical references and index.
    ISBN 1–57392–785–6 (paper : alk. paper)
    1. Animals—Therapeutic use. 2. Pets—Therapeutic use. I. Title.

RM931.A65 G73 1999
615.5—dc21   99–056545
                                                                      CIP

Printed in the United States of America on acid-free paper

CREATURE COMFORT

Dedicated to the memory of
Lea and Israel Gross,
and
to Lizzie, Lucy and Stephen,
so young, yet so brave.

CONTENTS

ACKNOWLEDGEMENTS

This book would not have been possible without the support and co-operation of the following organisations and individuals. Heartfelt thanks to:

The Society of Companion Animal Studies; The Children in Hospital and Animal Therapy Association; The staff and residents of The Royal Star & Garter Home for Disabled Ex-Service Men and Women; Birnbeck Housing Association; Dolphin Human Therapy Florida, USA; Dolphin Reef, Eilat, Israel; Dolphin Plus, Florida, USA; Conny Land, Switzerland; International Dolphin Watch; Canine Partners for Independence; Pets As Therapy and their volunteers; Guide Dogs for the Blind Association; Dogs for the Disabled; Hearing Dogs for Deaf People; Support Dogs; Chenny Troupe, Chicago, USA; The Riding for the Disabled Association; The Fortune Centre; Green Chimneys, New York, USA; Ealing Riding School; The Elisabeth Svendsen Trust; The Turkish Tourist Office; Panel Trend.

Dr Hyman Gross; Patsy Willis; Olivia Nuttall; Laura Graham; Sandra Stone; Ronnie Stone; Helen Cottington and Carrie; Nina Bondarenko; Dr Anthony Podberscek; Anne Docherty; Mary Whyham; Dr June McNicholas; Marion Jacobs; Julia McAvoy and Liam; Dr David Nathanson; Chris Connell; Sister Chiara Hatton-Hall; Pat Morrell; Sue Gaywood; Angela Howard; Robert Shillingford MBE; John Silverman; Jo Bomford; Sue Markham; Cathy Feiner; Val Neff; James Kelly; Julie Courtney; Dr Jo Burrill; Dr Martin Roiser; Dr Clare Twigger-Ross; Elisabeth Palfalvi;

Kurt; Donny De Castro; Mike Jones; Alexandra Murrell; Mary and Theresa Jones; Sarah Mathews; Alex Turner; Chris and Caroline Hykiel; Pauline Hume; Ann McBride; Rupert Sheldrake.

Sincere thanks to the dolphin therapy families: Geraldine and Natasha Graham; Alan, Caroline and Hayley Miles; David, Ailsa and Craig Sutherland; Tony, Dawn and Jamie Warwick. Your co-operation was invaluable.

To all the other animal species that help us humans – thank you, we are in your debt. I would like to express my particular appreciation to Biba, the PAT Dog, without whom much of my work would have been impossible. To Bebe my cat, and Sam my best dog friend, special strokes for the love and support that has kept me sane over the years.

To Ali Gunn at Curtis Brown, and Nick Webb, Helen Gummer, Ingrid Connell and Rochelle Venables at Simon & Schuster. Thank you for your support and encouragement – I couldn't have done it without you.

PREFACE

This is a book for everyone about the potential health benefits that can be derived from the company of a wide range of animals. However, the thoughts and desires of a psychologist are rarely straightforward, so if the following remarks initially appear to contradict this overall objective, please bear with me.

As a child, I was denied many of the normal animal experiences enjoyed by others because of my mother and sister's asthma. Furry animals made them exceedingly ill, which, for your average kid growing up in the 60s, left just goldfish and tortoises as possible pet companions. Tortoises didn't really hold much appeal for a six-year-old, especially as all I really wanted was a dog. However, I decided to give goldfish a go and very quickly became attached to Pixie and Dixie, who were acquired from the local pet store. Like many children, I took my responsibilities for feeding and caring for them very seriously, but unfortunately my lack of environmental awareness resulted in the most tragic consequences.

I kept the fish in a standard goldfish bowl and when this required cleaning I transferred them temporarily into a washing-up bowl. On one very sunny afternoon when I was changing their water, I thought Pixie and Dixie would enjoy some fresh air, so I placed their washing-up bowl holiday home on a chair in the back garden. I then went back into the kitchen and completed my cleaning activities, which took no more than five or ten minutes. Once I had finished I returned to the garden just in time to see a black cat's tail disappearing over our back fence. At first nothing registered in my young mind, but as I went to collect the washing-up bowl, the full horror resulting from the cat's brief visit became all too apparent. All that was left of my charges were two little heads floating on the surface of the water. For ten minutes I sobbed uncontrollably as I experienced my first bereavement but my father's offer of an ice cream did much to ease my feelings of loss.

Even though I loved the company of friends' dogs, I remained petless until I left home. It wasn't until I was living in Sydney in 1981 that, rather unexpectedly, and albeit briefly, I took in a dog. I had made my home with some friends in a four-storey apartment block five minutes' walk from Bondi beach. We shared a back veranda with a friendly Maori couple and their children. Quite early one morning their youngest daughter, who couldn't have been more than six years old, knocked on our door and presented me with the most beautiful black Labrador puppy. She explained that she had found it but her mum had said she wasn't allowed to keep the very young 'girl dog'. She was very keen for me to have her because then she would still be able to see her. I agreed to take the pup but I made it very clear to her that I would probably have to find her a new home because our apartments weren't really suitable for keeping dogs in. However, she seemed pleased that the dog would be around for at least a little longer.

It took me about three minutes to fall for the beautiful animal, whom I called Martha, but I knew deep down that ours was to be a very short acquaintance. I only kept her for two weeks and I spent most of this time trying to locate her owners but with no joy. In the end I walked the length and breadth of Bondi asking all the shopkeepers if they could give the pup a home.

Eventually I found a very well-to-do hairdresser who was more than willing to care for her. He asked if I wanted anything in exchange but I made it clear to him that the only thing I wanted was for her to be well looked after. He accepted this, but just I was about to leave his salon he called out, 'You could do with a trim'. So one of the strangest deals of my life was struck as I swapped Martha for a haircut.

It was still several more years before I had my very own pet and I really do believe that it was this early deprivation that fed my interest and need to explore human/animal interactions. As a mental health worker for many years, I was immediately drawn to the therapeutic potential of our association with other species, a subject that has received an increasing amount of media attention in recent years. However, the use of animals to complement more traditional medical treatments is by no means a new development. In ninth-century Belgium, animals were used in Gheel as a means to help restore the spirits of people with handicaps and disabilities. The first recorded therapeutic use of animals in Britain was in the 1790s. The York Retreat offered an alternative method of caring for people with severe mental health problems that contrasted quite dramatically with the inhuman treatment more commonly applied at this time. Patients at the Retreat were treated with kindness and respect, and encouraged to care for rabbits and poultry. It was thought that patients could learn self control by nurturing creatures weaker than themselves. Some 80 years later therapeutic horse riding was introduced in France for people with neurological disorders to promote balance and enhanced motor abilities.

The 20th century has seen many developments in this fascinating field and this is the stuff that the following pages are made of. I have attempted to offer a realistic warts and all overview of the work of the practitioners, researchers and animals involved in animal assisted therapy (AAT). My commentary on these laudable endeavours is presented through descriptions of my own experiences, research findings and a collection of personal accounts and case histories. This book is very much intended for the general reader but I do hope academic observers and specialists in AAT and related fields will afford me the honour of their company through at least some of its pages. I have done my best to adopt a balanced approach to some of the more contentious issues regarding animal captivity and the like, but none of us are without our prejudices and convictions so I've tried not to pull my punches.

Above all I've attempted to make this reading experience as pleasurable as possible. Please be advised, however, that given the nature of the activities described herein, there will be points along the way where tears may fall.

Bernie Graham
September 1998

PREFACE TO THE AMERICAN EDITION

As a 'Brit' sitting down to write a preface for the American edition of *Creature Comfort*, I can't escape the feeling that this activity is laced with a little irony. This is primarily due to the fact that so much of the content of the following pages is drawn from American spheres of endeavour. To be precise, the majority of the activities from the world of Animal Assisted Therapy described herein, whether in the realms of research or practice, emanate from across the 'pond'. A number of the most notable pioneers of Animal Assisted Therapy in the twentieth century, including Boris Levinson and Sam Ross, are sons of the Stars and Stripes.

Boris Levinson, a psychotherapist, was the first to document a pet animal's potential to ease children's communication in hospital or treatment settings in the early 1950s. His first observations were drawn from a chance event when a very withdrawn child he was treating arrived early for an appointment. Levinson's dog, Jingles, happened to be in the office with him and it quickly became apparent that the child, who interacted in a very positive way with the dog, was less distant and more inclined to communicate. Jingles in some way acted as an intermediary by breaking the ice and enabled the child to feel more at ease in the treatment environment.

Sam Ross is the founder of Green Chimneys in New York. This highly regarded Animal Assisted Therapy project offers residential care to children with special needs mostly from New York and the surrounding area. Many of the children have histories of neglect or sexual, physical, or emotional abuse. A good number also have deficits in their education as a result. The guiding principles at Green Chimneys are that through their work with a range of domestic, barnyard, and rare animals, children begin to learn responsibility and by nurturing the animals they may begin to nurture themselves.

The American influence on my written offerings is most clearly apparent in two of the major contributions to *Creature Comfort*. The first comes in the shape of Dr Hyman Gross, who acted as my medical mentor throughout this project. He lives in Los Angeles and works as a neurologist from his practice in the rather lovely Marina Del Rey area. His support has been invaluable and has ranged from funding my research with PAT (Pets As Therapy) dogs, to answering questions on the dynamics of blood pressure measurement at four in the morning! And how do I repay him for his help and guidance? I spell his name incorrectly in the first edition of the book published in England. This was a rather embarrassing mistake on my part because with the name Hyman, one should be particularly careful with its spelling!

For the second of these special contributions we must traverse the American continent to the beautiful Florida Keys. Here we find Dr David Nathanson at the Dolphin Human Therapy project. His openness and honesty about the potential and limits of his work are commendable, as was his facilitation of my interviews with families attending the project.

While speaking of David Nathanson, this may be an appropriate place for a news update. At the end of my chapter describing David's work, I informed the reader that he was about to move his operation to Mexico. This he did for a short while but is now back in Florida and working with Jeanie and Alfonso, a pair of dolphins who are rather special to my family and me. You will hear a lot more about them later on. The British climate hardly provides the ideal backdrop for dolphin encounters whereas the Florida environment is almost perfect. I sought the company of these wonderful creatures in both settings and I think the American reader may draw some comfort from my descriptions of these contrasting experiences.

In *Creature Comfort* I have drawn from the work of individuals in a number of countries and from differing backgrounds. However, one could argue, the resulting book is essentially a cross-fertilisation of British and American thought and enterprise. It is particularly interesting to note that both of the good doctors mentioned above were born in Britain and emigrated to the United States as young children with parents.

Following an eminently sensible suggestion from Steven Mitchell, my American editor, I have created a separate contact address section for American organisations towards the end of the book. This aside, the text is unaltered.

As I sit in Kew, in Southwest London, on a very rainy and chilly day, it is all too easy to drift back to the Keys. I do hope it is not too long before I return.

Bernie Graham
January 2000

HOW TO USE THIS BOOK

My overwhelming desire is, of course, that you would read this offering from cover to cover without pausing for sleep or sustenance. My ego aside, it may be appropriate for those considering implementing AAT programmes to review as necessary the final part of the book, which focuses on the practicalities of such ventures and also provides contact details of AAT groups and related organisations.

I personally find the range of theories offered on the 'whys and wherefores' of AAT quite fascinating. For some, however, this may not be such a treat. If this is the case for you, please be assured that I've made every attempt to make this material as accessible as possible.

However you use this book – enjoy!

PART I

SOME FIRST THOUGHTS

CHAPTER 1

THE CURRENT STATE OF PLAY

There is nothing quite like the old-style British mental hospital. The often tranquil rural settings cannot compensate for the intimidating Victorian institutional architecture. Even though over the last decade I've visited many clients who have found themselves both voluntary and involuntary recipients of care on draughty wards in such places, my feeling of unease on entering these surroundings remains undiminished. However, on one particular occasion I was a little less apprehensive. This was primarily due to the presence of Max, a six-month-old black Labrador puppy, who had placed his head inches from mine as he eagerly viewed the outside world through the front window of the Volvo. His surprisingly pleasant breath had been caressing my right cheek for most of the 30-minute journey to Napsbury Hospital on the outskirts of northwest London. He had also treated me to the occasional ear licking, which I would like to report as a supportive response to my increasing anxiety but this would be deceitful, since Max was most undiscriminating in what he licked. In fact as we drove through the hospital gates I suddenly remembered the last part of his own anatomy he had licked before me.

Max's 'owner', Julia McAvoy, and I were visiting Liam, a resident of the mental health housing project we were both involved in. He had been in hospital for several weeks suffering from a severe bout of depression. Liam was especially fond of Max and had looked after him on several occasions. We had informed him the day before that we were bringing the dog up for a visit and it was clear from his response that he was very excited about seeing him again. It was to be Max's first visit to the hospital and, given his extremely lively nature, we thought it best he didn't enter the wards. The plan was that while Julia visited another client, Liam and I would take him for a walk.

As we progressed further into the hospital grounds following the directions to Larch ward, the grotesquely ornate water tower came into view. This was a particularly intimidating feature of the hospital. It was no more than 45 metres (150 feet) high but it had the presence of a structure twice its size. Its styling appeared to be an unfortunate hybrid of Gothic and what I can only describe as Babylonian features. It was as though a beacon had been erected, to proclaim to all who entered the hospital, the indisputable dominance of the Victorian institution.

By the time we eventually arrived at the ward, Max was itching to get out and explore. It was initially necessary to keep him on a leash in case other patients and staff were less inclined to make his acquaintance. I stayed with him while Julia went to collect Liam. When she returned there was quite a reunion. Max and Liam were very pleased to see each other. After five to ten minutes of jumping, licks and hugs, Liam and I took Max for a run in the extensive grounds. We headed for a more secluded area so we could take the straining canine off the leash. We chatted in between throwing a stick and playing tug of war with Max. Liam explained that he was feeling a little more positive and was hoping to return to the housing project in the very near future. We also talked about football and the less-than-delightful hospital food. Throughout our conversation Liam rarely took his eyes off Max and smiled more readily than I had seen him do for some time. We spent an hour or so together before returning to the ward to meet Julia.

Max reluctantly got in the car and Liam said his final

farewells to him through the window. As we got in the car, Liam gave Max a toothy grin. Julia said it was good to see him smile again. He chuckled and told us that a nurse had also noticed him smiling earlier in the day and he recounted their conversation:

'In all the weeks that you've been on the ward I have never seen you smile first thing in the morning – it's good to see.'

'Well today is different,' Liam responded.

'Why?' the nurse enquired.

'Well today Max is coming to see me,' said Liam, smiling again.

Liam returned to the housing project within a month of our visit and is doing well. He has a job placement and is studying at a local college. His relationship with Max captures the essence of animal assisted therapy (AAT) – whereby interaction with an animal helps to foster human health and well-being or enhances quality of life. We all have some familiarity with the practical/functional support certain animals provide, guide dogs for the blind, for example, but AAT has much to offer in relation to the physical, mental and emotional aspects of health. The affect can be derived from the continued presence of an animal through both 'pet ownership' and contact made through visits to or from animals.

You will have noticed that I have placed 'pet ownership' in inverted commas. This is due to my strongly held belief, and please excuse me if I'm preaching to the converted, that due to the mutually supportive nature of our relationship with many animals, the term is both inaccurate and patronising. For those of you who have not seen the light, one of the objectives of this book is to bring you into the flock, as well as offer an overview of the many AAT programmes throughout the world and their effectiveness and provide some practical guidelines for those wishing to embrace AAT.

I must clearly state from the outset that I do not believe AAT to be a universal panacea. However, I do think that it has a significant contribution to make within a holistic approach to health, whereby it is the person as a whole who is treated, their individual physical, mental, emotional and spiritual needs being addressed where appropriate.

Research Overview

A report on AAT in the USA revealed that in 1977 there were 15 animal therapy programmes being investigated by 8 university research projects. By 1982 this had grown to 75 and 44 respectively. In 1998 the Internet search engine 'Yahoo' offered over 1,000 matches when 'animal assisted therapy' was entered. Interest has grown from predominantly academic and specialist origins to a more public fascination with the therapeutic potential of animals. Presented below is a selection of research findings relating to the effects of AAT.

Physical Health

In the late 70s and early 80s, a group of American researchers studied the survival rates of 92 people treated for myocardial infarction (heart attack) and angina pectoris. A one-year follow-up indicated that significantly more survivors owned pets. Fifty out of 53 pet owners were still alive in comparison to only 17 out of 39 non-owners. This finding remained significant even when researchers allowed for the beneficial effects of regular exercise on those who walk dogs. Similarly, when the researchers considered how the severity of the participants' heart conditions may have affected the outcome, they still found pet ownership to be significant in predicting survival.

An Australian study carried out at the Baker Medical Research Institute, Melbourne, in 1992 strongly indicated that companion animals are associated with a reduction in known risk factors for heart disease. Five thousand people aged 20–60 attended a cardiac risk evaluation clinic and completed questionnaires on their diet, exercise and family history of heart disease. They also had their blood pressure measured and a blood sample taken. They were then asked if they had a pet. The results revealed that male pet owners had significantly reduced cholesterol, blood fat and blood pressure levels compared to non-pet owners. These findings could not be explained by other factors such as smoking, diet or socio-economic status. In women, where it is acknowledged that cardiac risk factors only increase at menopause, it was found that pet owning women aged 40–59 also benefited from significantly lower blood pressure than women who did not have pets. These findings were irrespective of type of pet owned and once again could not be explained on the basis of smoking, weight, exercise or socio-economic status. Even more startling was the finding that pet-owners were found to have higher meat, fast food and even alcohol intakes than the non-owners in the study

Several studies have identified the stroking of an animal as an effective means of relieving stress and reducing blood pressure. I believed this effect to be confined to animals with fur or hair until I witnessed a very stressed stockbroker neighbour unwind one evening by stroking his pet chameleon.

Mental Health

1984 was an extremely busy year for AAT research. It was at this time that a major supporter of AAT, Clark M. Brickel, conducted his own inquiry into the effects of introducing a pet dog into the individual psychotherapy sessions of a group of depressed elderly people. Fifteen subjects were randomly assigned to three groups. One group received individual animal assisted psychotherapy where Fudge, a female dachshund, attended each therapy session. The dog was available to be held, stroked or talked to by each client. At the end of each session clients were requested to watch the dog for a short period. Members of the other groups received conventional psychotherapy or no treatment at all. Depression ratings were assessed before and after the study which lasted four weeks. Average levels of depression in the conventional psychotherapy group fell by 6.4 points. The levels in the AAT group fell by 11.6 points.

An earlier enquiry by Francis, Turner and Johnson studied the effects of visiting puppies and kittens on depressed, chronically ill adults. These patients were divided into two groups; one was visited by puppies and kittens, the other by humans (known and unknown). Significant improvements were found on depression rating scales for the group who had had access to the pets. Those only visited by humans showed no improvement.

In 1986 Alan Beck and his colleagues investigated the impact of birds on therapy and activity groups for two matched groups of eight and nine psychiatric patients diagnosed as suffering from schizophrenia. The patients had been in hospital on average between three and five years. Daily sessions of the groups were held for ten weeks in identical rooms except for the presence of caged finches in one of them. The patients were evaluated before and after the sessions using standard psychiatric rating scales. The group of eight who attended their therapy and activity sessions in the room which contained a cage with four finches had significantly better attendance and participation. They had also significantly improved in areas assessed by the Brief Psychiatric Rating Scale relating to hostility and suspiciousness. It was originally planned that the study should last for twelve weeks but it was concluded after ten weeks due to an unexpected development: four out of the eight patients in the 'bird group' were no longer in the hospital. Three had been discharged and one was on extended leave. All the patients from the other group were still in hospital. Were the birds in some way instrumental in this exodus? This is difficult to assess. Nevertheless, the coincidence is intriguing, and Scully and Mulder have certainly got out of bed for less!

In 1984 Katcher, Segal and Beck found that when dental patients about to undergo an extraction watched fish in a small aquarium, they appeared more relaxed but not to a significant degree in statistical terms. This hardly surprises me since I'm absolutely sure that no creature on earth could reduce my anxiety before having a tooth pulled!

A more recent piece of Japanese research studied the effects of stroking horses on three groups of people, one of which was made up of individuals with negative attitudes towards companion animals. The findings of the study suggested that participants experienced a decrease in tension after stroking the horses and that the horses also seemed to enjoy the encounter.

General Health

Judy Yates studied the perceived value of a visiting pet therapy programme to 7,500 residents in 70 nursing homes. The views of both nursing home directors and the volunteers who accompanied the animals on their visits to the nursing homes regarding the perceived benefits of the programme were found to be very similar. Both groups agreed that the most highly rated benefits related to a general area described as 'improved quality of life'.

In 1989 attendance was analysed at a major US city in-patient psychiatric unit over the course of two years. It was found that an AAT group attracted the highest percentage of in-patients voluntarily choosing to attend an occupational therapy group. It also became apparent that this group was the most effective of all in attracting isolated individuals regardless of diagnosis. The authors also concluded that AAT was an effective tool for diagnostic observation and assessment.

An English study on the health implications of pet ownership was conducted by James Serpell in 1990, at the Companion Animal Research Group, Cambridge University. He studied changes in general health following the acquisition of a pet. Participants were interviewed just prior to getting a pet and at intervals for ten months afterwards. Having a pet resulted in measurable improvements in general health over the period of the study. The general health of those who had acquired a pet was significantly better than that of the non-pet owning control group. The findings could not be explained by an increase in walking amongst new dog owners, because a significant, though less marked effect, was also found in cat owners

Bucking the Trend

It must be said that several studies have produced contradictory findings or found no evidence to suggest that interaction with animals is of therapeutic value: Lago, Connell and Knight (1983) found no apparent indication that pets had any impact on the physical health, mortality, social activity and well being of 55 elderly subjects.

Robb (1983) interviewed 37 elderly people on psycho-social and health related issues and found higher levels of morale amongst pet owners. However, Robb and Stegman (1983) replicated the study with 56 people but found no significant differences between owners and non-owners. In a more extensive survey in 1984 Lawton, Moss and Moles examined 3,996 elderly persons and found no association between pet ownership and well-being or health.

These mixed research findings on the benefits of pet ownership seem to indicate that the promotion of a simple beneficial relationship between pet ownership and health may be inappropriate. The human/animal bond would appear more complex and health benefits of ownership may depend on a range of factors. June McNicholas and her colleagues at Warwick University advocate that maybe the reason some of the discussed studies did not find an association between pet ownership and enhanced physical and mental well-being, was due to the timing of the investigation. They believe pets provide social support to owners and it is this support that enhances well-being. They suggest that social support is only visible when it is needed i.e. when it is mobilised at times of stress. Thus studies which investigate health and/or well-being over shorter time periods, or in the absence of stressful events, are less likely to find differences between pet owners and non-owners.

CHAPTER 2

A PERSONAL VIEW

Ibegan working in the mental health field in August 1987, in a 20-bed community based unit in west London. A supervisor was assigned to steer me through my induction period and ease the startling transition from the theory based academic study of mental health to the practicalities of day-to-day working with the 'mentally ill'. Staff at the unit worked a shift system, enabling 24-hour cover. This involved working several on-call/sleep-in shifts through the night each week. Many late evenings were spent discussing the nature and treatment of mental illness. On one such evening my supervisor mentioned that she had recently read an article advocating the therapeutic effects of swimming with dolphins. Such ideas were indeed thought-provoking, but a scream in the night soon relegated any further consideration to the subconscious. Nevertheless, a seed had been sown that was to germinate four years later.

One Saturday morning, early in September 1991, I was watching television with my then four-year-old daughter, Laura. We were watching a BBC children's' programme called *The Eight-Fifteen From Manchester*. An item was introduced about a young boy who had recently lost a parent. He had been invited to Amble, a small town on the Northumberland coast, to swim with Freddie, a hermit dolphin who had made his home in the waters nearby. The narrator explained that it was believed by some that 'sad and depressed people' could be 'cheered up' by swimming with wild dolphins.

The young boy, Daniel, was taken out to sea in a fishing boat and escorted into the water by a diver and Dr Horace Dobbs of the International Dolphin Watch

(IDW). The film showed Freddie immediately approach Daniel and they began swimming and playing together. Freddie allowed Daniel to stroke his underbelly and make other physical contact. When Daniel returned to the boat, he was asked how he felt. He replied, 'I know it sounds stupid, but I feel lighter.' It was at this point that I sensed the tears in my eyes and realised that Laura, beside me, was crying. When the item finished, the programme returned live to the studio in Manchester. The studio was full of children who, before the film, had been loud, excited and vibrant. Now the silence was deafening. I wanted to know more. The seed was beginning to grow.

Within two weeks I was on my way to Amble to meet Freddie the dolphin. My first enquiries had led me to Pat Morell. She co-ordinated trips to meet Freddie on behalf of IDW, an organisation concerned with dolphin conservation. Pat had been featured in the BBC film and had swum with Freddie on many occasions. She quickly explained that the only way, in her opinion, for me to gain any perspective on Dolphin Therapy was to swim with a dolphin myself. 'Could you make it next weekend?' she asked. It was now late September and she explained that this was likely to be the last trip before winter. Even though somewhat unprepared, I thought being literally thrown in at the deep end might prove to be a valuable experience, so I agreed to go. Unfortunately, I managed to pick up a quite severe chest infection in the week prior to my visit and I therefore arrived in Amble with a hacking cough, running a high temperature and feeling totally unprepared to jump into one of the world's most polluted seas.

I met up with Pat, her daughter Ruth and six others who, for a variety of reasons, were also keen to meet

Freddie. Pat later explained that some of the party were suffering from 'mental distress'. We were all booked into a guest house close to the harbour, which was regularly used by those visiting Freddie. The owner informed us on our arrival that Freddie appeared to have been injured by the blades of a boat's propeller. She explained that he enjoyed having a 'jacuzzi' in the wash of boats but somehow, on this occasion, he had got too close and had suffered some deep lacerations. Local fishermen were concerned to keep visits to a minimum to allow him time to recover from his injuries. As a consequence we agreed to make only one trip.

Once aboard the open fishing boat, I changed into a wet suit. Given I had a swelling North Sea with a 25-mile-an-hour gusting wind for company, this was no mean feat for someone uninitiated in the ways of the diving fraternity. We were approximately half a mile off shore when Freddie appeared for the first time. He sprang from the murky depths about 3 metres (10 feet) from our boat. I was stunned by his size and bulbous body shape. This was no sleek smiling Flipper, this was a quarter of a ton of seawise hermit dolphin – a bloody whale. I began to feel a marked sense of unease and the adrenaline started to pump. The boat was beginning to feel like a dentist's waiting room.

I attempted to calm myself by going over the instructions in the quayside notices describing the appropriate and safe behaviour to be observed if one made close contact with Freddie. These were summarised into several bullet points. But I could only remember point three, which informed us that 'it was not uncommon for a male dolphin to extend his usually retracted penis as a sign of friendship . . .' and even then I couldn't recall whether one was supposed to reciprocate.

Freddie began to swim under and around the boat. The first of our group went into the water but he paid little attention to them. We were all becoming increasingly concerned because the injuries were clearly visible and we agreed to be particularly careful if we made close contact with him. However, Freddie did not seem too interested in making contact and it was necessary for us to follow him in the boat to remain in close proximity.

I was becoming quite uncomfortable in my wet suit. The tightness of fit was accentuating the breathing difficulties I was experiencing, due to my condition. It was at this point that Pat Morell turned to me and said, 'Your turn'. My heart sank and my anxiety levels soared. However, I slowly entered the water. I soon found myself bobbing around at the mercy of the currents. The water was cold and I was beginning to experience quite frightening difficulties with my breathing. However, I decided to remain in the water as long as I possibly could. I was starting to feel a little more at ease, when the diver accompanying us announced that there was a large jelly fish in the water and that we should keep well out of its way because it had a particularly nasty sting. That was it for me. I was out of the water within 30 seconds. The rest of the group continued to enter the water but Freddie, even though he surfaced nearby several times, made no close contact.

From my discussions with some others in the group, it appeared that they felt they had derived positive effects from their time in Amble, even though no close contact with Freddie had been made. This may highlight other possible therapeutic aspects of the Amble experience; hydrotherapy, the relaxing effects of being in water; positive social interaction; experiencing a new and predominantly natural environment. In addition, the expectation of an out-of-the-ordinary encounter may have contributed to the positive effect. Such an experience may also offer a distraction from the mundane and day-to-day existence. Contact with Freddie, even from some distance, may indeed have provided the focal point of a powerful and restorative environmental experience.

Even though my attempt at close contact with the dolphin was unsuccessful, I was able to spend considerable time with Pat Morell. She described how some people she had accompanied, who were suffering from depression, had derived a healing effect from their contact with Freddie. She was, however, particularly critical of those who made unfounded claims of universal effect and cure. Pat also expressed concern for Freddie's welfare. She informed me that he didn't take to everyone. He had been known to 'nip' one or two disrespectful individuals and on one occasion had chased someone out of the water.

I left Amble with mixed emotions. My visit had been thought provoking and overall thoroughly enjoyable, but my dip in the North Sea had been alarmingly anxiety inducing. Maybe if I had met Freddie my feelings would have been different. (This was indeed the case when, some years on, I came face to face with two dolphins and swam with them in a lagoon in the Florida Keys – a tale I will recount a little further on in this chapter.) Nevertheless my time in Amble with Pat convinced me that I should conduct my own study. However, her comments echoed in my mind when several months later I asked a group consisting of 'schizophrenics', 'manic depressives' and 'clinically anxious' clients, 'Would you like to meet some dolphins?'

The decision to conduct my own research presented me with a host of practical difficulties. My immediate thoughts were to arrange a trip for a number of clients from the mental health unit where I was working at the time to swim with Freddie. Unfortunately, while I was involved in the initial preparations for the trip, it was announced in the media that Freddie had disappeared. The closest alternative was Dingle Bay in Ireland, but this was way beyond my financial means. No funding was available from any other source, so I was left in something of a dilemma; should I abandon the research or continue by attempting access to dolphins in captivity? I had, and still do have, mixed feelings about dolphins being kept in captivity. On the one hand, I believe that these creatures should be free, not restricted in comparatively small pens and made to perform for the public. However, I felt the opportunities for contact between dolphins and humans had done much to raise public awareness, enhance positive attitudes to marine conservation and enable research for the benefit of dolphins in the wild. In addition, I certainly had reservations about plunging several vulnerable clients into even the most inviting of waters after my own experience. Their welfare was my paramount consideration and even though a highly experienced psychiatric nurse, Alex, had agreed to accompany me on any research trips with clients, I very quickly came to the conclusion that contact with dolphins in a more controlled situation was desirable.

I decided to approach Windsor Safari Park to ascertain if research could be carried out there. I was invited to meet with Steve Walton, who had overall responsibility for the dolphins at Windsor. Steve met me outside the dolphin arena and, after introducing himself, asked me a number of questions about my proposed research. He indicated to me that I had answered the questions to his satisfaction and then invited me into the empty arena to meet the dolphins. Once inside he apologised for the grilling and explained his motives. He informed me that he received many requests from researchers and 'the like' for special time with the dolphins. A recent experience, however, had made him very wary of these approaches. He had, some months before, agreed to a seemingly bona fide 'researcher's' request to meet the dolphins. However, on entering the arena the individual concerned immediately climbed the small stadium's stairs and stood on one of the uppermost seats, faced the dolphins, closed his eyes, placed his hands on his temples and pronounced 'I have made contact, I am now communing with the dolphins.' Steve didn't go into the fine detail of how this episode was concluded but it would be enough to say that the 'communing' individual was quickly escorted away.

Steve took me to the side of the pool and it was here that I had my first close contact with a dolphin. A very sleek smooth skinned male bottlenose dolphin swam to us, raised its head, looked at Steve and made those wonderful clicking and chuckling noises so easily understood by the park ranger and his sons in *Flipper*. I naively enquired what the dolphin was saying. I regretted my utterance before the last syllable left my mouth. Steve gave me a sidelong look and replied, 'He's probably saying hello.' He returned the greeting and then put his hand in the dolphin's mouth and tickled his tongue. The dolphin gently moved his head from side to side appearing to thoroughly enjoy the sensation Steve encouraged me to do the same. I carefully placed my hand in the creature's mouth, fully acquainting myself with each and every one of his sharp conical teeth, and tickled the tip of his tongue. It felt very similar to a human tongue. I can categorically state that I have never trusted any animal so readily. I've put my hand in few dogs' mouths that I have had an established relationship with, when their owners had wanted to demonstrate the dogs' temperament (or my gullibility), but never before on a first date.

The dolphin remained close by as we discussed the practicalities of my future visits. From the corner of my eye I noticed him filling his mouth with water which he then proceeded to flip over Steve. Finding this most amusing, I laughed out loud at which point the dolphin repeated the exercise, this time at my expense, before swimming away to join two of his peers at the other end of the pool. My second encounter with a dolphin had proved to be a very pleasurable and uplifting experience. I left Windsor a little damp but in extremely high spirits. I was completely unaware of how short-lived my state of mild euphoria was to be.

Steve Walton had agreed to my clients watching the dolphins' performance and then having some private, individual time with them, allowing supervised close contact from the pool side. I set about selecting subjects for the research project and made all the necessary arrangements. All seemed to be going well, when it was suddenly announced that, due to financial difficulties, Windsor Safari Park was to close immediately. I telephoned Steve at the first opportunity and he very apologetically informed me that he had been instructed by the Receivers to cancel all planned research. I did wonder at this point whether my study would ever come to fruition. However, Steve put me in touch with Peter Bloom at Flamingo Land in North Yorkshire, the last remaining dolphinarium in the country. He was most accommodating and arrangements were soon made for me to take a group of clients on a 450-mile day trip to meet Lotty, Betty and Sharky – the dolphins at Flamingo Land.

The trip to meet the dolphins was approved and financially supported by both the local social services department and the community mental health team. However, just prior to our departure, one of the local psychiatrists began to vocalise objections to my research. He claimed the experience could be detrimental to the clients' health, leaving himself and other medical staff to 'pick up the pieces'. He did not substantiate his comments and since all the clients participating in the study were very willing volunteers who seemed to primarily view the trip as a day out, his views were without any foundation whatsoever. He was, thankfully, overruled by a more senior consultant.

The objecting doctor's comments did, however, compound my feelings of frustration with the dominance of drug therapies, and the resistance to the use of complementary or alternative approaches in the treatment of mental health problems. My own work, and the experiences of my clients, have convinced me that the adoption of an holistic approach is the way forward. This involves the use of client-centred therapeutic approaches to address individual needs that go beyond broad, quantifiable biochemical treatments. I would never deny the value of the appropriate use of drug therapies but the 'appropriate use' is often difficult to determine. These decisions are in the main made by psychiatrists, many of whom, I believe, are fixated on drug therapies as the only treatment of real value and advocate that we must live with the side-effects and dependencies which often occur. The difficulties in establishing such approaches are profuse. The past prejudice of many, but not all, in mainstream

psychiatry against psychotherapy as a valid interven-
tion has been profound. So a venomous reaction to my
research into pursuing an holistic approach was not
unexpected.

The day of the trip to Flamingo Land arrived and
we all gathered at the arranged meeting point at the
local railway station. The group consisted of myself,
Alex, the psychiatric nurse, and six clients – three men
and three women. In terms of formal diagnoses the
group make up was as follows: two of the women were
suffering from protracted depressions and the third
was labelled 'manic depressive'; one of the men was
'schizophrenic', another was very sensitively described
by his psychiatrist as being in an 'anxiety state' and the
third was recovering from a severe head injury and
trying to come to terms with the resultant changes in
his personality. As I viewed the assembled party and
pondered the logistics of the trip, it may have been
more accurate to have described me as the 'anxiety
state' of the party.

As we descended the stairs to the station platform
one of the female clients, Mary, informed me that she
had become extremely frightened of underground
train travel as a result of the horrendous fire at King's
Cross station which had claimed so many lives.
However, she added that even though she had not
used the London Underground since the fire, she was
prepared to give it a go if the journey wasn't too long
because she really wanted to see the dolphins. I
explained that the shortest route necessitated being
underground for about 10–15 minutes at the most.
Mary said that she thought she would be fine. I
encouraged her not to worry and told her I could do
some breathing exercises with her if she became
anxious. She seemed fairly relaxed as she boarded the

waiting train and took her seat. As the doors closed and the train began to pull out of the station, Mary turned to me and said, 'I know you went over the travel arrangements with us but I was having a bad time on the day, and I don't mean to be a pest but where do we pick up the main line train to Yorkshire?' I should have seen this coming but there was nothing I could do now to lessen the impact of my reply, 'King's Cross'. The colour instantly drained from her cheeks. I really thought at that point that the journey would end before we left London. However, even though Mary became very distressed, she insisted on continuing with the journey. I tried to find an alternative route but we had to travel from King's Cross. Her anxiety level was clearly very high when we got on the Underground. Alex and I did all we could to support her, but through her own determination she made it to King's Cross. On arrival in the underground part of the station, the actual scene of the fire, she took herself very quickly up to the main station area at ground level. We caught up with her as fast as we could and after ten minutes or so she appeared to be a little more at ease. She had been extremely brave and whatever happened now with the dolphins, making the trip had necessitated her confronting her fear – an achievement in itself. Thankfully the rest of the party were fine and the remainder of the train journey passed without incident.

We changed trains at York and headed for Malton, our final destination. Peter Bloom had agreed to collect us from the station in his minibus. There were more people in our group than Peter had expected and we only just squeezed into the vehicle. The 15-minute ride to Flamingo Land was pretty uncomfortable and Mary became very distressed again and by the time we reached the dolphinarium, she was in tears. I accompanied her as we entered the poolside area trying to offer some comfort. As the dolphins came into view, one of them swam towards us and raised her head seemingly to look at us and 'clicked-chuckled' at us. The change in Mary's emotional state was instantaneous. Her tears turned to laughter as she began calling to Lotty the dolphin. A few moments later I was sure I saw Mary skip a step. I wasn't sure if I'd imagined it so I asked Alex if she had noticed anything and she confirmed that she had seen it too.

Louise, who also suffered from a protracted depression, appeared quite anxious at the beginning of the trip too. However, as the day progressed this certainly seemed to diminish. She spent most of the journey chatting to one of the male clients with whom she was particularly friendly. Much to my amazement, she attempted a few dolphin impressions chorused by bouts of giggling. I had never seen her so relaxed before.

All in all we spent about two hours with Lotty, Betty and Sharky throwing balls which they flipped back with their beaks and playing with hoops. Peter Bloom took all of us individually to stroke and feed one of the dolphins. We all spent a little time viewing them underwater through large windows in the basement of the arena. Sheila, the women diagnosed as manic depressive, a very refined lady, particularly enjoyed watching the dolphins underwater and at one point pressed her face against the glass to compensate for taking her glasses off. It was at this precise moment that Betty the dolphin decided to deposit yesterday's lunch inches from Sheila's face. There was a cry of 'typical' as she turned towards me almost crying with laughter. She was not alone in her high spirits. All of the group clearly enjoyed their time with the dolphins.

Everyone in the group returned to London with a smile on their face. However, it wasn't until a week later that my analysis of the data collected before, during and after the trip, began to reveal some fascinating responses to our time with Lotty, Betty and Sharky. All six clients were asked to complete inventories evaluating their levels of anxiety on three occasions: a week before the trip, while in the company of the dolphins and a week after the trip. In addition, I conducted short individual interviews with the group in the week following our return, focusing on their reflections on meeting the dolphins. To supplement this data, Alex and myself had made separate observational notes during the trip.

I began analysing the data by extracting the 'before', 'during' and 'after' scores on the anxiety inventories. I double-checked the somewhat startling figures. They indicated that Mary's and Louise's levels of anxiety, which were the highest in the group a week before the trip, fell by over 20 per cent while they were with the dolphins. A pattern began to emerge suggesting that Mary and Louise, both diagnosed as 'depressed', had derived a similar effect from their time with the dolphins. In their interviews, they both described how much they enjoyed their time with the dolphins and their sadness on leaving them. They also reported feeling much calmer following their contact with the dolphins, an effect which lasted for several days after the trip. Louise also related in her interview that the last time she had felt this way was when she had seen some dolphins in a seaside display some years before.

I was fascinated by Mary's and Louise's similar responses, but for me the most intriguing effect, again derived by both of them, only came to light several weeks after the trip. A little background detail is required: even though I had known Louise for three years I had never been her 'key' or named worker. She had attended assertiveness groups I had led but I did not have a detailed knowledge of her past. With Mary on the other hand, I had a close professional relationship. I had only known her for a year or so but I had been her keyworker throughout that time. I met with her at least once a week for emotional support sessions and I was very aware of the many difficulties she had experienced in the past. As a child she had suffered horrendous sexual and psychological abuse at the hands of a man employed in her parents' business. She believed her ongoing depressed state was a direct result of her childhood experiences.

Mary attended a therapy group for women who had suffered sexual abuse as children, run by a local clinical psychologist. It was only by chance that I found out that Louise also attended the sessions. Mary and Louise could not be described as friends. They were certainly acquainted with each other but had little contact outside of the group. Following the first group session held after the trip to meet the dolphins, I received a phone call from the facilitating clinical psychologist. She explained that she had asked Mary and Louise for permission to contact me because of comments they had made regarding the dolphins. They had both related to the group that, in some way, the dolphins had enabled them to reflect on, and to some extent confront, the loss they had experienced as a result of their abuse as children. They could not specifically articulate how the dolphins had done this, or what the loss was, but it was suggested that the feelings of trust and innocence, engendered so readily by the dolphins, may indeed have been the major emotional casualties of their childhood experiences.

The findings of my study and the far more extensive work of others, have convinced me of the positive effects of human interaction with other animals. However, a chapter describing my personal perspective would not be complete without me recounting the experience of swimming with Jeanie, a female dolphin, and Alfonso, her four-year-old son, which I was pleased to share with Patsy, my wife, and Laura, my daughter, who by then was 11.

Ever since my dip in the less than inviting waters of the North Sea, I had longed to swim in the company of dolphins and hopefully enjoy the frequently described uplifting effects of such encounters. A last minute decision, at the end of a week of interviews for this book in Miami, took Patsy, Laura and myself to Dolphin Cove on Key Largo, just under fifty miles north of Key West. Our guides for the day were Will Gilbert, a retired Professor of Biology and avid bird watcher, and Nancy, his daughter, an animal trainer. A week earlier, they had given us the most enjoyable and informed tour of the Everglades. Nancy had made inquiries on our behalf regarding supervised programmes for swimming with dolphins and luckily, at very short notice, she managed to book three places at Dolphin Cove on the nine a.m. swim.

The night before in our hotel room Patsy, Laura and myself had discussed our feelings about the imminent adventure. Laura was very excited but to varying degrees the three of us shared similar apprehensions; would the dolphins take to us? Were we strong enough swimmers? Would we be swimming in very deep water? What else would be lurking in that deep water? Sharing our feelings went some way to reduce our levels of anxiety but we all went to sleep that night

feeling a little uncertain of what would unfold the following morning. During the night I dreamt I had been
sandwiched between two pieces of bread and was
about to be consumed by an irate sea creature. I awoke
in a sweat trying to work out what kind of creature it
was and what action to take. The vivid sandwich sym-
bolism seemed self explanatory but another possible
interpretation will be discussed a little later on.

Will and Nancy picked us up at our hotel in Key
Biscayne at 6.30 a.m. and we headed for Dolphin Cove
down the US 1 highway. The Keys were beautiful in
the hazy morning light but I found it difficult to focus
on the surrounding scenery because my trepidation
was too much of a distraction. Memories of my North
Sea experience were drifting back and forth. I tried to
persuade myself that this was a completely different
situation but it's at times like these that your mind
takes on a 'mind of its own'. It took us just over an
hour and a half to reach the cove. We were very early
so we went for some coffee nearby. I made sure I had
decaff because caffeine was the last drug I required.
The time soon arrived to return for our dolphin
encounter.

Dolphin Cove is a deep-water lagoon on Florida Bay
which is home to four dolphins – Spunky, Duke, Jeanie
and Alfonso. On the bank of the lagoon there is small
centre offering facilities to strictly limited numbers of
people who wish to swim with the dolphins or scuba
dive in the surrounding waters. The grounds are very
well kept with tailored emerald lawns and white grav-
elled driveways. It seemed to be a quite new centre and
one building was being completed at the time of our
visit. On entry we were asked to wait in the allotted
area for our 'trainer' but we couldn't resist a peak at
the dolphins who were surface diving and jumping out

of the water. The lagoon's entrance from the bay was closed off by a metre (3-foot) high wire fence and I was informed by the staff that this was there to keep people out rather than to stop the dolphins leaving. We were assured that they had the ability to leave if they wished and Nancy confirmed that this certainly seemed the case from what she could see. To an untrained eye the dolphins certainly seemed happy to stay – maybe it was the regular food which meant there was more time to have fun.

A group of about 12 people had now gathered in the waiting area and just on nine o'clock we were welcomed to the centre by one of the trainers who invited us all to join her for a brief boat ride in the bay for half an hour's tuition. We were informed that this instruction session usually took place on dry land but some manatee had been spotted close by earlier in the week and if we were lucky we might be able to see them. None were seen but we were treated to an educational and rather entertaining half-hour talk on the physiology of dolphins utilising 'Flipper', a fluffy stuffed toy, as a demonstration model. We were also instructed on how we should behave in the water, informed of some 'dos' and 'don'ts', and were encouraged to maintain eye contact with the trainer as much as possible. Our instructor assured us that the dolphins were very gentle and that we had nothing to fear. She told us that these highly trained dolphins had been used in therapy programmes with little children and how she was far more concerned about our behaviour towards them. Finally, we were shown how to use buoyancy jackets and it was explained that our encounter would be in two parts. Our first meeting would be at the quay-side in pairs, where we would get to know the dolphins a little by rubbing and stroking them. This, we were told, was also to give the dolphins a chance to size us up.

There were eight people going to swim with the dolphins in our session and we were divided into two groups of four. My group consisted of myself, Patsy, Laura and Gale, a friendly lady from Canada. Before too long it was time to make our way down to a floating wooden platform to meet Jeanie and Alfonso. We were to be introduced in pairs and we had decided amongst ourselves that Laura and I should go first. The sun was now breaking through, but the beads of sweat beginning to form on my forehead were more as a result of being told that the spinach green water I was about to enter was over 4.5 metres (15 feet) deep and home to a population of 'snapper' fish. Laura and I knelt on the edge of the platform with Lynn, our trainer, between us and in no time at all two smiling faces broke the surface of the water. Jeanie and Alfonso were before us and, under the careful watch of Lynn, we began to gently stroke these enchanting creatures. Their skin was very smooth and to the touch their body mass seemed to have the consistency of a shelled hard boiled egg. Jeanie tilted her head and for just a moment she appeared to be gazing directly into my now moistening eyes. Her fleeting acknowledgement moved me – there was no fear now, just an overwhelming desire to join her in the water. I looked over at Laura, her senses seemed to be feasting on Alfonso's attentions. We spent a few more moments on the platform getting to know our hosts and then, very reluctantly, moved aside to allow Patsy and Gale to take their turn.

I turned to Laura and asked if she was OK. She said 'This is wonderful but I feel I like crying.' I told her that I felt very much the same. We also shared an eagerness to get in the water and meet our dolphin hosts on their own terms. Patsy and Gale soon finished their introductions and it was time for the first in our number to take the plunge. We were to go in one at a time and then in pairs. We asked Gale if she would like to go first since my family was dominating the group. She happily accepted the offer and lowered herself off the platform and into the depths. We watched as the mother and son team eased her through the water by pushing with their beaks against her feet. They repeated this manoeuvre several times with Gale on her back and doing a 'flying Superman' on her front. She was also treated to a dorsal fin tow. Jeanie and Alfonso swam into parallel positions either side of her and, on Lynn's instruction, she took hold of their back fins. Once her grip had been sensed, they took off at some speed. They completed a circuit of the lagoon and then deposited Gale at the side of the floating plat-form. I could barely contain the urge to leap in and join them and, thankfully, I didn't have to wait too long. Lynn called out 'OK that's it for now' and Gale climbed out, a little hesitantly, but she was assured that she would be spending some more time with Jeanie and Alfonso later on. Then the words were spoken – 'You're up Bernie.'

Now, with no reservations at all, I entered the lagoon. The water felt charged by the presence of the dolphins. I was grinning inside and out – the child in me was coming out to play. Anticipation somehow fused with expectation and as I lay on my back and felt their touch on my feet, a rush of reassuring calm claimed both my body and mind. This intense but oh so gentle euphoria was naturally complemented by the caress of the cool water. This sensation remained with me throughout my time immersed with Alfonso and

Jeanie. My dorsal fin ride was akin, in my mind, to a trip on a sea-bound magic carpet but this was real. The controlled power of their bodies rhythmically undulating as they propelled themselves and me through the water had a musical quality which I felt very privileged to share. Holding on to their fins I could feel almost every note but all too quickly the piece was over and I was hauling myself on to the dock

Laura needed no beckoning to enter the water. She was now face to face with Alfonso and her smile easily matched his. I was still glowing as I watched them from the platform. I had heard many people speak of dolphins' particular liking for children and now I was seeing it for myself. Laura's movement seemed more matched to that of the dolphins as she was taken for her rides. This may have been partly due to the fact that, if Gale would forgive me, she and myself could lose a few pounds. In fact when they were pushing me I could swear at one point I heard Alfonso panting. Laura grinned continuously throughout her time in the water. Due to her smaller body size, Alfonso and Jeanie seemed more able or inclined to take her for longer excursions and they had to be called back more than once when they reached the lagoon's entrance. Laura was a strong swimmer and I had complete trust in the dolphins to take care of her. This in itself intrigued me: I have never entrusted Laura's safety to a human so readily. Her time was now up and with great reluctance she climbed on to the floating dock.

Patsy was last in and somewhat gingerly she entered the water. Patsy is not the strongest swimmer and never before had she swum in such deep water. It was clear that, not unlike myself in the North Sea, she was a little frightened as she left the dock. In addition, she had expressed some concern regarding us intruding in the dolphins' environment. However, she put her fears aside and joined Alfonso and Jeanie. Unfortunately, she experienced some problems with her buoyancy jacket which resulted in her swallowing several mouthfuls of sea water. This, as Patsy later pointed out on her return to dry land, was of course my fault because I had put too much air in it! To her credit she persevered even though she continued to feel uneasy and not a little queasy. It seemed that Jeanie sensed Patsy's anxiety and tried to help. When it was time for her dorsal fin ride, Lynn encouraged her to hold on to only one of the dolphin's fins. As Patsy did this and held on for dear life to Alfonso, Jeanie placed herself under Patsy's body and supported her as she was pulled through the water. A little more relaxed and rather moved by Jeanie's support and feminine empathy, Patsy swam to the floating dock. This was not to be her day however, because as she climbed out she cut her thigh on a jagged edge.

It was now time for Laura and myself to swim together. Mother and son meet father and daughter. This was the second most wonderful event I have shared with Laura, the first being her birth. We did much the same together with the dolphins as we had done alone, but the sharing of this experience, especially the dorsal fin ride, with me holding on to Jeanie and Laura with Alfonso, will be treasured always. The finale to our time in the water was the two of us holding a wooden bar above our heads which Jeanie and Alfonso leapt out of the water to jump, first individually and then in unison. As we pulled ourselves on to the floating platform, the thought struck me how I had begun teaching Laura to swim several years ago at a very early age and now, with the help of the dolphins, it seemed her tuition was complete.

Due to the cut on her leg, Patsy decided against returning to the lagoon and, after expressing a little sympathy for her mum, Laura inquired if she could take her place in the pair with Gale. Lynn agreed and in she went again, relishing every moment. I waited on the dock and watched as Laura enjoyed this unexpected opportunity.

All things must come to an end but as Gale was climbing out, Laura was still playing with Alfonso close to the entrance of the lagoon. Lynn called to Alfonso and I beckoned Laura to return. Our words, it seemed, had fallen on deaf ears. They continued to play seemingly oblivious to our attempts to attract their attention. Eventually, after a minute or two they responded and swam to the dock. They then said their goodbyes. Laura and I left, with me gently and with only a tinge of seriousness, telling her off for continuing to play when it was time to come in. As we entered the changing area I could hear Lynn having a very similar conversation with Alfonso.

CHAPTER 3

THEORIES, THOUGHTS AND FEELINGS

Before I present some of the theories offered to explain the therapeutic benefits derived from contact with animals, I think it would be appropriate to consider our relationship with them in more general terms.

There are broadly two schools of thought on how our attitudes towards other animal species have developed. The first draws from a Jungian psychoanalytic approach. This suggests that the essence of our relationship with animals is innate. Through millions of years, so the argument goes, we have evolved and developed a predisposition to relate positively to certain species. This is based on the idea that millions of years ago humans learned that certain species could alert them to danger by their behaviour – dogs barking being the most obvious example – or enable them to avoid starvation by helping to track game. Negative/positive relations with certain species became part of a 'collective consciousness' which is passed down the generations internally, as it were.

The second view advocates that our responses to animals are primarily learned as we grow up within a particular culture or multi-cultural environment. From a very early age, children are encouraged to view certain animals in a positive or negative light and, depending on your culture, whether a species is considered a 'food' or 'non-food'. In vegetarian societies, children are obviously taught that no animal is considered a 'food'.

These views are reflected and reinforced in children's literature, songs, games, television, films and via word of mouth. Whether they are accurate or not appears of secondary consideration. This anomaly in itself lends a good deal of weight to the 'learned' camps' argument. As a young child, I was read 'Little Red Riding Hood' and the 'Three Little Pigs', where wolves were definitely the bad guys. I was also taught songs such as 'Who's Afraid of the Big Bad Wolf?' and played games like 'What's the Time Mr Wolf?'. Even though at cub scouts I was told the story of a family of wolves who had sensitively raised an orphan 'man child' in the jungle (do they have wolves in the jungle?), there was still no question about it – wolves were bad news. Bears on the other hand were presented to me and my contemporaries as beautiful, gentle and kind creatures. An endless stream of bear stories and characters filled our books and cartoons – 'Goldilocks and the Three Bears', Winnie-the-Pooh, Paddington Bear, Yogi Bear and Boo Boo, Barney Bear and Rupert Bear to name but a few. In addition, we were all given teddy bears to cuddle at night. In adulthood, however, we found out that the contrary is true; bears are to be feared because brown and polar bears can be extremely aggressive and life threatening, while experience and new research has shown that wolves very rarely bother humans. Even so, most people still believe that wolves are to be feared. If our fear of wolves had evolved from experience over millions of years, one would expect it to be well founded.

Personally, I lean towards the 'learned' camp. If we inherited our attitudes and were predisposed to react in certain ways to particular creatures, we would find it extremely difficult to change these views, feelings and resultant behaviours. If we consider the marked change of opinion that has taken place over the last 25 years relating to hunting, animal husbandry and the eating of flesh, it is extremely difficult to accept that millions of years of consciousness conditioning can be altered to such a degree and in some respects actually reversed, so quickly. I would also add that people's wide ranging views on animals would indicate a learning process derived from individual experience rather than a collectively inherited predisposition. Similarly, the innate view leaves little room for free will and choice. The learned behaviour argument also gains credence from a cross-cultural perspective; the British and the North Americans see dogs as beloved companions, whereas to several far eastern populations, a dog is often something to be eaten. Such a differing view of a species abundant in both the East and West is, to my mind, culturally conditioned.

Animals as Food

The consumption of certain animals within a culture/society appears to be primarily determined by nutritional need, availability and economic considerations. If financial conditions change for the better however, consumption is often continued out of (taught) custom and tradition. Religious belief is also a major influencing factor. Christianity, Islam and Judaism all advocate man's dominion over the rest of the animal kingdom. For many followers of these major world religions, this dominion has encouraged moral responsibility and sensitivity to the plight of the animal kingdom. However, for others within our societies, dominion has been equated with permission to torture, maltreat and exploit.

There are an ever growing number of people in society who no longer consume meat on moral grounds. For many, the killing of animals for food in itself is acceptable, but the modern-day intensive factory farming methods of meat production are not. The use of animals for food is often justified by the relationship between species in the wild: animals kill each other for food and this is seen as part of the natural process. I would argue that all these differing views should be respected, but I cannot condone the hunting of endangered species, cruel farming and transport methods, the killing of animals for 'sport' or 'fashion' and the torture of animals in ritualised practices to symbolise human dominion over these often defenceless creatures.

I think it only right and proper, in the context of my discussion of differing cultural views of animals, that I take this opportunity to offer a long-overdue apology to the good people of south Florida. One evening during my visit to the Dolphin Human Therapy programme, Patsy, Laura and myself decided to visit the Sonesta, one of the local hotels, for dinner. I had spent the day observing therapy sessions and interviewing the parents of children being treated while Patsy and Laura had enjoyed the marvellous weather sunning themselves by the pool. We were all in good spirits and ravenous. Selecting a restaurant or menu that suited us all was not without its complexities due to our various diets. Laura has been a vegetarian since she was eight years old and I then only ate white meat and fish. Patsy eats anything but doesn't mix certain foods together. On this particular occasion we were sure the chosen hotel could accommodate our range of needs since it offered three restaurants. We eagerly entered the lobby more than ready to eat. Having decided to check all three restaurants before selecting our final choice, we began to peruse the menu of the least formal, café type eating area. Within seconds my eyes focused feverishly on one of the dishes on offer – 'Cajun Dolphin Sandwich:

locally caught blackened dolphin between two pieces of the bread of your choice'. Never before have I reacted with such venom and hostility to a menu item. I wanted to give vent to my feelings but it was as though I was being choked by the numbing irony, that these caring creatures I had witnessed, only hours before, giving so much so readily to humanity were one of tonight's 'Specials'. It was not long before Laura noticed the offending item. She vocalised her feelings with no hesitation, 'Daddy they eat dolphin!' The tears were welling in her eyes. Patsy seemed in a similar state of shock to my own, but she tried to comfort Laura. I asked a slightly puzzled member of the waiting staff if they really served dolphin. 'Yes, sir! Freshly caught today.' His seemingly cold detachment did nothing to ease the situation. I was now ready to blow. Laura asked, 'Are you going to complain or say something?' Patsy was trying to calm us both down. I realised I must make a protest of some sort. So true to my English heritage, I took a deep breath, put on a stern expression, instructed the family to follow my every action, then sighed as loudly as possible and left the hotel. We regrouped in the car park and I explained that, at the last moment, I felt that, as guests in a foreign country, we should respect the local culture, however disgusted by it we may be. My excuse fooled no one, least of all myself. Our outrage subsided a little as we made our way to a nearby Italian restaurant we had noticed earlier in the week. We had decided to try and dissipate our upset by having a slap-up Italian meal. We didn't check the menu this time and on entering we were escorted to a table by a very friendly Spanish waiter. Patsy and I ordered two stiff gin and tonics to calm ourselves. When they arrived however, it quickly became apparent that our order had been misunderstood. We slowly began to make our way through two ferociously large undiluted gins. We were going to say something but on reflection after a couple of sips, decided we would suffer them. Laura also seemed refreshed and a little happier after a couple of mouthfuls of Coke. Then it happened again. We looked at the menu and there it was, 'Grilled Dolphin'. This time I didn't hesitate. I called the waiter over and exclaimed, 'You serve dolphin!' 'Yes, sir,' he replied. 'I am very sorry, though, because we have none today.' Before I could respond he added, 'I go see when we have it next time', and quickly disappeared. Within a few moments, however, he was back at our table. He paused for moment and placed his hand very gently on my shoulder. He then thoughtfully declared, 'Sir, I think my answer is not the one you wanted, there was something in your eyes.' I told him he was very perceptive and explained that we were very surprised to see that dolphin was caught for food. He then assumed a demeanour that clearly conveyed that this was not the first time he'd had this conversation. 'This dolphin we eat is not porpoise dolphin flipper. This is another fish – mahee mahee. Very good to eat. We cannot catch or eat porpoise dolphin flipper. The law no allow this.' We enthusiastically thanked him for his explanation and shared a communal sigh of relief. I did, however, begin to feel a little guilty about some of the thoughts that had rushed through my head that evening, regarding the local people and culture, but a couple more 'misunderstood' gins helped eased my conscience.

I mentioned this episode to Chris Connell, the public relations officer for Dolphin Human Therapy. She confirmed that we were certainly not the first British visitors to suffer this confusion. As a result, an explanation of this rather puzzling use of 'dolphin' is now included in the therapy programme.

Thoughts on Theory

Many theories have been proposed to explain the therapeutic effects of human interactions with companion animals. I offer but a small, and hopefully palatable, selection below. For 'dessert', I will consider some of the suggested explanations for the positive effects derived from encounters with dolphins. Theories range from those suggesting that companion animals offer comfort, social support and distraction, to those advocating that contact with animals enables human beings to re-engage with natural energies. Most of the theories share the views that *i)* pets may offer constant, non-judgmental, and often unconditional, love and respect to individuals who may be unable to derive this from their relationships with other humans; and *ii)* pets seem to have the capacity to make one feel needed, which may be crucial for the development, re-development and maintenance of self-esteem. This, I might add, could relate especially to older people who live alone following the death of their partner. This is readily understood as we have all experienced loneliness at some time in our lives.

Social Settings and Social Theories

These theories advocate that human-animal relationships are influenced by differing cultures and should be viewed within their social and environmental settings. They also assume that the psychological aspects of such relationships are essentially the same as those underlying person-person relationships. They do not in themselves offer explanations for the healing effect derived from companion animals, but suggest a framework to help understand our relationship with them, which may, in turn, offer some clues to their therapeutic potential.

Proponents believe that although social human-animal relationships may vary from exchanges between persons, these relationships can be viewed in terms of the 'person-environment fit'. The person-environment fit approach seeks to understand how an individual is affected by their living situation. Such evaluations focus on the physical, psychological, social and economic aspects of someone's life. Advocates of this approach believe that the psychological and social aspects of our lives are of particular importance to the understanding of human-animal bonding. These aspects include role performances, relationships, knowledge and understanding, order and control, and activity. Responsible pet ownership involves certain role performances and expectations (feeding, exercising, visits to the vet, etc) and can be characterised as a relationship requiring special knowledge and understanding. It may contribute to or disrupt order and control in one's environment and requires some care by the owner depending on the pet's needs.

Two theoretical frameworks have been suggested which may be helpful in understanding the roles and relationships pets play in a person's environment: the social role and exchange theories.

Social Role Theory

A role is defined as any set of behaviours that have some socially agreed upon function and for which there exists an accepted code of norms. Four dimensions characterise the effect of social roles on the individual: *1)* the number of roles; *2)* the intensity of involvement; *3)* the pattern of participation over time; and *4)* the degree of structure the role imposes. In our society, each stage of one's life may be characterised by varying roles that assume these four dimensions. For example, the 'normal' child has fewer roles than an adult but great intensity in these few roles, and a pattern of consistency if s/he dwells in a 'healthy' environment with a great deal of structure. A responsible role which is culturally accepted in our society for younger children is that of pet owner. Expectations surrounding the role of pet ownership are that s/he will develop a sense of responsibility, that it may provide a confidant for the growing child who listens and doesn't ask questions.

Laura, my 11-year-old daughter, is extremely responsible in her care of our cat Bebe and her goldfish, as many children are. She feeds them with no prompting and cleans the fish tank on a regular basis. She also walks Bruce, a bull terrier pup, for a local shop owner. She takes a pride in these activities. It's just a pity she doesn't feel quite the same about tidying her room! When upset, she is usually to be found in her (untidy) room stroking and talking to Bebe.

As a person ages, multiple roles develop – parent, worker, etc. Pet ownership may be one of many roles and, depending on the circumstances, may remain as important or become less significant. On reaching an advanced age, however, a person may experience role loss (through widowhood, retirement, illness, etc). Some persons rapidly replace lost roles with new ones, others rejoice in their new found freedom and others mourn their loss. Pet ownership may become a burden to older people who wish to travel, whereas it may be a more significant role for the more isolated. For the latter type of person, creating roles with other persons may be less of an option.

Exchange Theory

Exchange theory suggests that people continue to engage in relationships as long as the benefits outweigh the costs. Advocates of this approach highlight the fact that the literature on human-animal relationships has often focused on children, those with mental health problems, the disabled, the elderly and other special population groups. To these groups, they believe, pets may provide valuable relationships that serve such functions as companionship, tactile stimulation, safety and non-judgmental emotional support. Therefore the benefits may greatly outweigh the cost in certain situations.

I find that this theory explains little as millions of people from outside the 'special populations' cited have also been shown to derive a range of benefits from their companion animals.

Therapeutic Theories

James Serpell, a leading researcher in the field, believes that no all-embracing theory can account for the benefits derived from animal assisted therapy. He proposes a 'general effect' theory which suggests three ways in which animals may be of therapeutic value:

Instrumental

This approach relates to horse riding therapy and the use of guide, hearing and 'Pets as Therapy' (PAT) dogs for those with special needs. People who are mentally and physically challenged, and as a result lack self-esteem and confidence, are able to control an animal without such disabilities and use it as an extension of themselves. Serpell believes this can increase co-ordination, mobility and skill and hence improves confidence and self-esteem. I will consider this element of Serpell's theory in more detail in the chapters focusing on dogs and horses.

Passive

Passive interaction involves watching animals such as fish or birds. Serpell argues that becoming absorbed in such animal activity can induce a state of relaxation. The benefits are primarily short term and last as long as the animal is observed. He considers animals to be especially effective in this regard because (to humans) their activities are relatively random and unpredictable and therefore capable of sustaining interest. The animals are essentially objects in this situation and Serpell proposes that one could probably derive a similar but lesser benefit from a piece of kinetic sculpture or an open log fire. Personally, I agree with him up to a point, but I feel the difference between watching animals and kinetic sculptures is far more pronounced, and probably beyond comparison. Animals are organic life forms, with individual idiosyncrasies and a potential to interact with the observer. Ongoing relationships are often established between the observer and the observed. Humans affectionately nurture fish and birds they watch, whereas our relationships with fires and kinetic sculptures are rather less involved.

Anthropomorphic

This category relates to animals as pet companions with bonding potential to the owner – dogs, cats, etc. Serpell believes that a fundamental aspect of this category is that the therapeutic benefit depends on the person initially perceiving the animal as another person. Once this has happened, the behavioural signals transmitted by the animal are seen as expressing attachment, devotion and love to the person. He submits that psychological and physical well-being are dependent on an individual feeling respected, needed and loved by others. It would seem that animals can supply this in some cases when humans cannot. He states:

> 'Therefore, animal therapy of this kind would be expected to have most benefit for individuals who feel unloved, rejected, socially alienated or friendless.'

I think that includes just about all of us at one time or another in our lives. Serpell concludes that,

> 'The notion that all these different types of therapy could have the same general psychological and physical outcome is not as unlikely as it seems . . . A similar process is recognised as operating in the opposite direction in the case of the "general stress response." In this case we have a multiplicity of unpleasant stimuli, ranging from illness to divorce affecting people's mental and physical health in similar detrimental ways. It therefore seems reasonable to postulate that a variety of pleasant stimuli could induce equally "general" effects of a restorative nature.'

I certainly agree that companion animals have the ability to offer devotion, support and love to a person. However, I do not believe that one necessarily has to 'initially perceive' the animal as a another person. Quite the opposite. Over the last seven years, I have had my closest canine relationship with Sam, a lurcher/greyhound cross. I adore him because he is a loving, affectionate and supportive *dog*. Our friendship is simple and uncomplicated. He offers me a refuge from the complexities of human interactions.

Serpell's theory has much to offer to our understanding of the therapeutic role companion animals play. I believe he has certainly identified many of the essential components of the therapeutic effect and his work has prompted much of the recent work in this field.

June McNicholas and her colleagues at the University of Warwick advocate an analysis of the supportive aspects of our relationships with companion animals as a means to furthering our understanding of the therapeutic effects. They believe that pets may act as providers of social support. They define social support as the:

> 'Social, emotional and instrumental exchanges with which the individual is involved, having the . . . consequence that an individual sees him or herself as an object of continuing value in the eyes of significant others.'

They state that mainstream health psychology has shown the importance of social support in the recovery from a wide range of medical conditions as diverse as hip fracture, first stroke, major depression, osteoarthritis, heart attacks and cancer, as well as stressful life events such as bereavement of a human partner. Emotional support is seen as particularly valuable in the early stages of serious illness or a major stressful life event. June McNicholas puts forward three ways in which pets may function as providers of social support:

1) Pets are perceived as always available, predictable in their responses and non-judgmental. They provide a sense of esteem, being cared for and needed regardless of the owner's status as perceived by self or others. Personally, I think being needed is extremely important. Keeping a pet offers us a means to meet the very human need to care, look after or nurture, on very uncomplicated terms. This is especially so for people denied this – the childless maybe or the elderly who live alone, for example.

Pets can also give tactile comfort (from strokes and hugs), recreational distraction for owners from their own worries, and in addition are not subject to 'provider burn out'. Thus they provide a constant source of comfort irrespective of fluctuations in human support.

2) No social skills are required to obtain their attention, thus removing the potential problem of how to ask or elicit support. Individual social abilities in negotiating or regulating social support are not applicable, therefore avoiding mismatches in required/received support or apparent shortfalls in received support.

3) Pets may provide a 'refuge' from the strains of human interactions, allowing a freedom from pretences or barriers that may necessarily be erected between giver and recipient of support to mutually protect the relationship. This release from relational obligations may provide both a breathing space and an opportunity for 'naturalness' that has particular relaxation benefits. Pets also offer the opportunity for freer demonstrations of emotion, unrestricted by social convention.

Marie-Jose Enders-Slegers of the University of Utrecht also advocates theories relating to the social support functions of pet animals. In addition, she believes *i)* the relationship with a companion animal is such that it affords control and responsibility to the owner and fosters a sense of self worth and usefulness, and *ii)* companion animals provide stimulation in a person's environment, offering sensory experience and encouraging mobility. Much of her work has focused on the benefits of pet ownership to elderly people. The additional support aspects she identifies seem particularly relevant to older people who live alone and have minimal contact with other people. Just as a side issue, I think it is important to note that busy parents, with several children, who already have enough responsibility, may find a young puppy less than conducive to their well-being!

Pet Ownership and Visiting Programmes

Much of the discussed theory relates to pet ownership. However, certain elements can also be applied to visiting pet programmes or those situations where time with an animal is likened to a treatment. These include: the predictability of response (consistency) of a pet, the fact that they are non-judgmental, that no social skills are required to elicit their attention, and that they offer a short-term refuge from the strains of human relationships, allowing a freer expression of emotion/affection. The aspects of the proposed theories that are particularly relevant to visitation conditions relate to companion animals providing stimulation in an individual's environment through recreation, distraction, sensory experience (tactile and visual) and stimulation of mobility.

If social support is accepted as the major therapeutic agent in human-companion animal relationships, it may be argued that recipients of visitation pro-grammes, or short-term treatment programmes, receive lower and/or less sustained doses. However, my own studies and those of others, indicate that signifi-cant benefits may still be derived in these situations. Recipients of pet visitation/treatment programmes are in the main 'ill' or require therapy, and in the majority of cases, in contrast to pet owners, they interact with animals in non-domestic 'sterile' environments (hospi-tals and nursing homes, for example), but they never-theless still derive a therapeutic effect. Maybe, the effect of the contrast of a dog and a hospital ward is, in itself, a therapeutic environmental stimulant.

June McNicholas also cites the perception of pets as 'always available' and ready to provide a 'constant' source of comfort irrespective of fluctuations in human support as significant elements in the function of animals as providers of social support. These two elements are clearly not present when pets are part of a visiting programme or treatment, but again thera-peutic effects may still be derived.

One could argue that all of the above contribute to our understanding of the dynamics of AAT. My own personal belief is that the therapeutic process is determined by a combination of factors, but that the following are of particular importance: *i)* social support; *ii)* control; and *iii)* environmental change. The role of social support has been extensively described, but the other factors may require a little further explanation

Control

For many people it often seems as though they are at the mercy of life events or dependent on the actions of others. This can be especially so for those who are sick or disabled. Patients in hospitals or residents of nursing homes are 'looked after' and expected to be passive recipients of care. Disabled people are often treated, very unwisely I might add, as incapable of choosing how they would like to lead their lives. However, in our interactions with companion animals, the dynamics are somewhat altered. As Marie-Jose Enders-Slegers proposes, the relationship with a companion animal is such that it affords control and responsibility to the owner, which can foster a sense of self-worth and usefulness.

A large-scale study at University College London Medical School strongly suggested that feelings of being in 'control' at work are related to good physical and mental health. I would contest that these findings are also relevant to life in general, and I believe that certain visiting programmes have the potential to enhance self-esteem. With horse riding for the disabled, for example, children and adults who may have very limited mobility or may be confined to a wheelchair have the opportunity to learn to control the movements of another living creature.

Environmental Change

I've touched upon this notion in previous pages when describing the ability animals have to offer a distraction from sterile hospital surroundings and normalise these environments. In essence, they have the potential, by their presence, to dramatically change environments, whether they be care institutions or peoples' homes. The idea that environmental change in itself can be therapeutic is not a new one. Two common expressions come to mind: 'A change is as good as a rest' and 'I need a change of scenery'. Our surroundings can affect us physically, mentally, emotionally and spiritually. The addition of an animal could be seen as small dose of 'Environmental Therapy'. In terms of the physical, a different living element is added to the environment. Their presence is not a static one and as a result movements, interactions and activities in general may be affected. This animation can alter our perception of an environment. For some people, the acquisition of a pet simply brings company into a previously lonely house or flat.

Animals also have the ability to make situations more normal, usual, domestic or homelike and, above all, less threatening. This can enable people to feel more at ease with a clinical environment. This process is seen by many practitioners as being of significant therapeutic value. The term 'therapy' is often defined as 'that which heals' or 'to take care of'. However, its full meaning also includes 'that which makes whole'. Many psychologists promote normalisation as an essential aspect of therapeutic environments. They believe that the lack of ordinary/normal everyday experiences can detract from the therapeutic process and leave individuals feeling acutely inadequate. How can the individual be 'whole' if deprived of these ordinary/normal experiences? An obvious difficulty of this approach is the debate surrounding the nature of normality. That aside, the assertion that the presence of companion animals normalises hospital/care home environments derives support from several studies. In 1979 Clark M. Brickel undertook research to assess the therapeutic role of cats on a total care ward for the elderly. Patients were only able to see the cats during the day in the dayroom. Even though the cats lived in the hospital they were only allowed in the dayroom. They were fed and slept outside. Staff interviews indicated that the cats had been performing a socially beneficial role and had served as positive stimuli enhancing the patients' level of responsiveness. For some patients who began to develop an emotional attachment to the cats, there seemed to be a knock-on effect which enabled them to establish better relationships with peers and staff. In addition staff related that the cats contributed in a positive fashion to the ward environment. The presence of the cats was seen to cut into the institutional atmosphere of the ward, rendering it more home-like and pleasant.

In the 1986 bird study conducted by Alan Beck, I referred to earlier in Chapter 1, he compared the impact of therapy groups on two matched groups of psychiatric patients who met in identical rooms except for the presence in one of the rooms of caged finches. He found that the group which met in the room with the finches had a significantly better attendance and participation record. His findings indicated that this difference was due to the fact that the room with the finches was perceived to be less threatening and safer.

In 1992 research was conducted in Sweden into the use of integrated treatment programmes for 'severely disturbed psychiatric patients'. It concluded that when pets were allowed at centres – in an attempt to both enhance the family type atmosphere and to exploit the psychological benefit of having access to a companion animal – they did indeed humanise the atmosphere.

Several studies, including my own, have suggested that pet visitation may also bring a sense of normality to the institution by acting as a social bridge. In terms of community care, a dog gives a volunteer a reason to visit and so promotes involvement by community members in the hospital or care home. Similarly, the walking of dogs by residents of community based mental health facilities can help to normalise relations with neighbours through increased social interaction and communication due to the dog's presence, while at the same time encouraging the perception of normal activity.

June McNicholas and her colleagues at the University of Warwick, in their work on the supportive role pets may play for the bereaved, also identify a normalising role. They suggest that pets may help reorganise and re-establish routines and initiate social contact in a way that is unconnected with the bereavement. After all, one has to continue to walk the dog following a bereavement. In addition, a dog's potential for initiating social interaction is not diminished by the loss suffered by his or her human companion.

Studies by a number of researchers, including June McNicholas, have demonstrated that positive social interactions increase quite dramatically when people are accompanied by a dog when out walking. Similarly, and maybe more significantly, research on the west coast of America has clearly shown that both children and adults in wheelchairs also experience far more positive interaction and communication with others when they are accompanied by their assistance dogs. They appear to act as ice-breakers and help the able-bodied overcome those all too common but unnecessary feelings of awkwardness when confronted with a person who has a visible disability.

Natural Environments

Animals, I believe, change the living environment of many town and city dwellers by introducing an element of the natural environment into their urban surroundings. Indeed it has even been suggested that animals offer a means for the estranged human to reconnect with the natural universe. Some theorists believe this to be the essential component of any therapeutic effect.

The therapeutic potential of being exposed to features of the natural environment was demonstrated by a thought provoking study conducted by Richard Ulrich in an American city hospital. He compared the recovery of patients whose ward/room window looked out on to a brick wall with those who had a view of a cluster of trees. The pleasant sight of trees was associated with shorter hospital stays following operations, more positive evaluations in nurses' notes and a decreased use of analgesic pain killing medication.

An in-depth evaluation of the 'restorative' potential of change in one's environment was undertaken by Stephen Kaplan and his colleagues in the 70s and 80s. Their work focused on the effects of being immersed in the natural environment – often referred to as the 'wilderness experience'. Over an eight-year period, groups of people whose age ranged from 14 to 48, spent up to 11 days in an area of wilderness in the northern USA close to the Canadian border. Nine of the 11 days were spent together and the remaining two days camping alone. The groups consisted mainly of city dwellers from the Chicago area. Stephen Kaplan was particularly interested in the effects, if any, of this wilderness experience on an individual's personal development. Data was collected by questionnaires and the recordings in diaries. Questionnaires were completed at the beginning of the trip, several days into the experience and at the end of the 11 days. The diaries were kept throughout and for a further six weeks after group members returned to their normal environments. The researchers were interested both in how people adapted to the wilderness and how their degree of self-confidence, self-esteem, insight and ability to cope were affected by their experience. Overall, significant positive effects were found in all areas. The information collected from the diaries after participants had returned home suggested that these changes were sustained and that the wilderness experience had profoundly affected their lives.

On analysing the benefits of the wilderness experience, Stephen Kaplan began to consider a number of factors which he believed were not unique to wild environments. He also proposed that these factors operated in a variety of surroundings, from which comparable benefits could be derived. He referred to these settings as 'restorative environments'. He identified four influencing factors from his wilderness work and suggests that in any restorative environment one or more of these must be present to some degree:

Being Away

This requires being away from at least one of the aspects of a person's usual, everyday environment. Kaplan proposed three different forms to being away: *Retreat*, whereby a person might get away from intrusive distractions by, for example, going to a quiet rural area, although any quiet, secluded place with no telephone may serve just as well; *Putting aside the work one usually does*, thus escaping from what Kaplan describes as 'a particular content' and anything that reminds you of that 'content'; *Internal escape*, whereby one is able to take a rest from pursuing certain purposes and possibly from mental and emotional effort of any kind. 'Being away' may involve any or all three of these.

Fascination or Interest

Being away and bored is unlikely to be of any restorative value. If people are particularly interested or fascinated by an environment, it allows them to function without having to call upon their capacity for 'effortful' attention. They can rely on the interest inherent in the environment to guide their behaviour, making effortful attention unnecessary. Hence they can rest that component of their 'mental equipment' which is so susceptible to everyday pressures and stresses.

Coherence

This refers to the potential of an environment to seem like another world and to absorb one's 'imaginings' but at the same time be understandable. This type of environment may function as 'another world' but at the same time, it does not violate what we know, and what we believe about the way things work in the world. In essence, it is stimulating but not threatening to our beliefs.

Compatibility

The fourth and final factor is 'compatibility' across 'domains of human functioning'. Kaplan suggested that activities required in a restorative environment should fit well with people's inclinations. He described this as 'consonance' or harmony between the necessary and the desirable which engenders a sense of simplicity and peace of mind. In the wilderness such activities revolve around food, shelter, fire building and so on. Kaplan believes this compatibility to be crucial to the extent and quality of the benefit derived from an environment where the other three factors are present.

Kaplan's work may offer some insight into our understanding of how animals provide a restorative quality to our day-to-day surroundings. Companion animals in our urban living environments may enable us to 'be away' from the complexities of many human activities and relationships. My cat, Bebe, fascinates me and I can certainly say my resulting attentions are effortless. For the dog owner, attentions may not be regarded as effortless in relation to regular exercise. However, walking the dog may enable someone to 'be away' and as a consequence, derive a restorative effect. This could be further enhanced if there is a 'consonance' between the need to walk the dog and the desire to do so.

Dolphin Therapy

I would now like to consider some of the theories advanced to explain the healing abilities attributed to dolphins. Pat Morell, whom I met in Amble with Freddie the dolphin, believes that the therapeutic benefits of dolphin contact may be derived from the dolphins' electromagnetic field. Dolphins have a much larger and more powerful electromagnetic field than humans. This, she believes, stimulates mentally distressed individuals and may act as a catalyst for recovery for some people.

Dr Horace Dobbs, of the International Dolphin Watch, has claimed that close contact with dolphins in the wild has resulted in some quite startling recoveries from depression. Dobbs's work at sea has not been substantiated scientifically and, for this reason, is difficult to evaluate. He has proposed that face-to-face contact in the sea with wild dolphins may help restore harmony with nature. This in turn, he believes, enables 'Chi' or 'Qi' energy to flow more freely. This concept is drawn from Taoist (ancient Chinese) philosophy. Taoists perceive the universe as a living organism infused and permeated by the life force – a rhythmic, vibrational energy. Dobbs shares the Taoist view that if Qi energy becomes stifled, emotional difficulties ensue that may lead to health problems. In general terms he believes that the Western existence and sophistication distance us from nature and can lead to disharmony of mind, body and spirit.

Two journalists, Amanda Cochrane and Karena Callen, in an overview of dolphin therapy, identify six 'elements of healing' which, they believe, contribute to the therapeutic effects of close dolphin contact in the wild: *i)* the power of joy – being in the water with dol-

phins is immensely joyful and euphoric; *ii)* the love of
the dolphin – anecdotal reports indicate that dolphins
direct their attention to those in distress and impart
'unconditional love'; *iii)* emotional release – the power
of the experience enables a release of pent up emotion; *iv)* the energy connection – the 'Qi' energy theory
put forward by Dobbs; *v)* hydrotherapy – the relaxing,
stress reducing effect of being in water; *vi)* the healing
power of sound – the whistles and clicks dolphins produce are therapeutic, enabling re-balancing. I can
certainly attest to a euphoric reaction to meeting
Jeanie and Alfonso, and feeling energised in their
company.

These theories offer much to contemplate. Dr
Dobbs, Amanda Cochrane and Karena Callen, share
the view that therapeutic benefits can only be derived
from contact with dolphins in the wild. However, it is
interesting to note that both my own research and the
far more extensive work conducted by Dr David
Nathanson, a leading authority on dolphin therapy,
indicate to the contrary.

My own view of dolphin therapy is that it provides a
powerful natural stimulant in an out of the ordinary
(restorative) environment. This may 'lift' some depressed people and by virtue of this poignant and
unusual experience in itself, they may begin to believe
that life can change (for the better). Having worked
closely over the years with many depressed individuals, I must stress that I do not believe dolphin
therapy to be effective for everyone struggling with
this condition. Differing attitudes and personality
traits must also be considered, as I learnt very quickly
when I was making the preparations for my own study.
Garth, a chronically depressed young man in his mid
twenties, seemed an extremely suitable subject for the

research, someone who I felt might derive benefit from meeting the dolphins. I invited him into my office and explained the details of the journey and visit. He sat there not saying a word. When I had finished he seemed disappointed. I asked if he was OK. He said, 'I'm all right, I'm OK, the train journey sounds great, but I'm sorry I don't like dolphins.'

In many ways my view of dolphin therapy echoes David Nathanson's assertion that it provides a 'jump start' for the children he treats, the rationale being that if the desire to interact with the dolphins is sufficiently strong, the child will focus long enough to give a correct response. As a result, the child must increase his/her attention span. He firmly believes that many of the children's learning difficulties are related to problems maintaining attention rather than their ability to mentally process information. Upon returning home to their regular therapy, this enhanced concentration level allows for more information to be processed, which results in accelerated learning. Dolphin therapy is a complement to traditional approaches not a replacement.

More recent theories suggest that dolphins' sonar may play a therapeutic role. Early research on the 'sonochemical' effects of dolphin sonar on human tissue and electroencephalographic results following human-dolphin interaction indicate increased human alpha wave production. Alpha rhythms are often associated with higher relaxed states.

Well these are the theories. It would now seem appropriate to consider dolphin therapy in practice.

PART II

PETS IN PRACTICE

CHAPTER 4

DOCTOR
DOLPHIN

I am a child of the 60s and was born into the *Flipper* generation. From the moment I first saw the dolphins in this very successful TV show, I fell in love with these wonderfully sensitive and intelligent creatures. As you may have gathered already, little has changed. However, before I begin my review of the application of dolphin therapy, I must subordinate fantasy to reality: I believe that the therapeutic use of dolphins has enormous potential but I do not endorse the view that these sea-born therapists are mystical or ethereal purveyors of healing. Similarly, I must stress again my belief that while dolphin therapy may be effective for *some* sufferers of a range of physical and emotional problems it is not a universal panacea. Unfortunately, there are some within the field of dolphin therapy who promote exaggerated claims regarding the healing potential of dolphins, and raise false hope and expectation amongst the most vulnerable. From the outset of the following exploration, I wish to distance myself from such individuals. I will take a closer look at the ethical implications of dolphin therapy a little later on.

We must also respect these mammals' own sensitivities, and to illustrate this point, a cautionary tale. Dolphins, in the main, seem to welcome the company of humans but on rare occasions they have been known to be less than pleased with our attentions. One of the most celebrated such incidents involved the multi-talented Robin Williams. In the mid 90s he made an excellent TV documentary on dolphins which included footage of several moving encounters he had swimming with them in the wild. His very first encounter, however, seemed more likely to move his bowels than his emotions. He entered the water with a little trepidation from a beautiful beach in the Bahamas to interact with two bottlenose dolphins who, he had been informed, were well used to swimming with visiting humans. The dolphins swam around him as he stroked their bodies and began to relax a little in their company. Seemingly from nowhere, one of the dolphins, with some venom, rammed him in the chest with his beak and very aggressively jabbed him a couple of times. The incident was over in seconds and the dolphin swam away. The ram seemed to be a reaction to being touched in a place where physical contact was unwelcome. I must stress that such experiences are the exception rather than the rule and in most cases they can be prevented if people are instructed in the appropriate behaviour to be adopted when encountering a dolphin. It is intriguing to note that young children seem to be excluded from such reactions.

When I began researching this field of therapy in the early 90s, some of the first work I came across described encounters between dolphins and young people suffering from autism. This pioneering work was conducted by Dr Betsy Smith in the Florida Keys. Autism, the developmental disorder which isolates the child or adult from the world as we see it, is a complex condition which, to a great extent, remains resistant to treatment, although advances have been made in its management with appropriate support. Believed to be caused by brain dysfunction, it affects children from birth or infancy. The condition varies in severity but impairs the natural instinct to relate to fellow human beings. Words, gestures and facial expressions can mean little to someone with autism. They show little curiosity or imagination and frequently seem indifferent to the usual process of two-way communications. As a consequence of these difficulties, people suffering with autism can have extremely limited attention spans which restrict their learning potential. Dr Betsy Smith found that interaction with dolphins can lead to consistent increases in attention span during and after each child–dolphin encounter. She reported that the attention span of some autistic children increased from 5–10 minutes to up to an hour. One of her most startling findings however, involved an 18-year-old young man who had been diagnosed as a 'non-verbal autistic child' at the age of six. After spending time with the dolphins, he began making dolphin 'clicks'. His clicks, when listened to on tape, were very difficult to distinguish from some of those made by the dolphins used in the study!

Dolphin Human Therapy

In the world of dolphin therapy all roads will eventually lead you to Dr David Nathanson. He uses a behaviour modification programme which emphasises the reward role dolphins may play for children with a range of conditions and disabilities.

Dr David Nathanson

Dr David Nathanson began pilot studies into dolphin assisted therapy in 1978 at 'Ocean World' in Fort Lauderdale, Florida. He then developed a two-day-per-week, one-therapist programme at the 'Dolphin Research Center' in Grassy Key, Florida, from 1988 to 1994. During 1995 and 1996, 'dolphin human therapy' (DHT) became a full-time project employing additional therapists at 'Dolphins Plus', Key Largo. The current therapy centre is located at the Miami Seaquarium on Virginia Key, but is soon to move to Mexico. More than 15,000 therapy sessions have been conducted since 1988, with over 800 children from 39 countries.

Dr Nathanson and his team work mostly with children with a range of special needs. Many of the children have multiple diagnoses. The most common diagnoses are cerebral palsy, Down's syndrome and autism. The theory/concept behind dolphin human therapy (DHT) is that children will increase attention if they can earn a meaningful reward, the reward being interaction with the dolphins. Depending on their condition, children are asked to perform tasks challenging them in such areas as speech and language, and motor skills. When they respond appropriately, they 'participate in behaviour' with the dolphins. Dr Nathanson and his colleagues are fully aware that ongoing dolphin assisted therapy is impractical in the vast majority of cases. However, they believe that the dolphins act in a motivating capacity and serve as a 'jump start' for the children, the rationale being that if the desire to interact with the dolphins is sufficiently strong, the child will focus long enough to give a correct response. As a result, the child must increase his/her attention span. Upon returning home to their regular therapy, this enhanced concentration level allows for more information to be processed, which results in accelerated learning. DHT is a complement to traditional therapy not a replacement. Dr Nathanson emphasises that DHT does not prevent or cure disease or disability. However, recent research indicates that DHT, by significantly increasing attention and motivation, can substantially reduce the amount of time required for children with multiple diagnoses to improve mentally, physically or behaviourally. Dr Nathanson's research also indicates that using DHT to quickly move children with severe disabilities to the next level of functioning appears to be cost efficient and practical. Indeed, his research strongly suggests that two weeks of DHT can achieve the same or better results than six months of conventional physical or speech therapy. As a consequence, in proportional terms, a significant financial saving is made. In our age of homage to cost effectiveness, this is a major consideration. His most recent research also indicates that in about 50 per cent of cases, gains made by children in a fortnight of DHT are maintained or improved after one year away from the programme.

Dr Nathanson's findings make for persuasive reading, but for me, seeing is believing and I needed to know more. So with the full co-operation of Dr Nathanson, in late April of 1998, I travelled with my family to the Florida Keys to see DHT in action. Just prior to our departure, two stories emanating from the DHT programme were given considerable media coverage and further inflamed my interest. The first related to Nikki Bryce, an eight-year-old boy from Weston-Super-Mare. At birth Nikki had been deprived of oxygen and, even though he was physically capable of talking, he had never spoken a word. After just three days of DHT he spoke for the first time. After being taken out of the water he pointed back at the dolphins and said 'in' – his first ever word.

Another young boy, nine-year-old Travis Matthews from London, who suffered from hydrocephalus (an abnormal accumulation of water on the brain) and mild cerebral palsy, could only walk when wearing callipers. However, after only one morning's DHT, his supports were removed and he walked alone and unaided by any apparatus for the first time.

Chris Connell, the DHT public relations person invited me to spend as much time as I wished observing therapy sessions and Dr Nathanson readily agreed to an interview. In addition, I very much wanted to gain some insight into the views and experiences of the families with children receiving therapy. I decided to spend a couple of days getting a feel for the programme and establishing a rapport with the families before asking too many questions.

All in all I observed seven morning and afternoon sessions of DHT over a four-day period. Each session lasted 40 minutes and comprised four children receiving individual treatments. Parents and other family members are welcome to watch the sessions but, due to insurance requirements, only the child receiving therapy, a dolphin trainer and the therapist are allowed to enter the water. Each child usually attended for one session of therapy a day, either in the morning or afternoon. During these sessions I introduced myself to the families and explained my presence. Purely by coincidence, all the children undergoing DHT that week were from Britain. Two were from Scotland, one from Northern Ireland and three from England. Of the six, four families consented to being interviewed. The other families expressed a little reluctance, and not wanting to pressurise anybody, I decided against pursuing the matter. Before I reflect on these families' experiences, I would like to consider, in a little more detail, the organisation, practicalities and processes of the DHT programme.

The therapy takes place in Flipper stadium. This is the smaller of two show arenas which comprise the

Miami Seaquarium on Virginia Key. The larger stadium is home to a performing orca ('killer whale'). Tiered seating is built around three sides of a roughly 60-metre (70-yard) square lagoon which is fenced off from the ocean. There are holding areas for the dolphins attached to the main swimming area. The lagoon is home to eight dolphins. Therapy takes place in the main swimming area which is up to 4.5 metres (15 feet) deep. A soft sandy bottom is visible through the very inviting crystal clear water. The stadium is built on and around the disused set of the *Flipper* TV series. In one of the open corners there is a small building, housing storage and changing facilities, with a sign over the door reading 'Ranger Station' – I was in my element. I'd come home.

Four floating platforms, measuring approximately 2.5 by 1.5 metres (8 feet by 5), are located in the corners of the main swimming area. It is on these platforms or stations that the therapy takes place. Each child, dressed in a wet suit, is taken to one of the stations for their session. They are accompanied by the dolphin therapist, a dolphin trainer and an intern who is undergoing training with the DHT programme. The trainer is responsible for supervising the dolphin's involvement. A senior trainer keeps a watchful eye on these very playful creatures before and after interactions with the children and acts as the dolphin traffic controller. They also seemed to be monitoring the dolphins' moods and assigned the dolphin who was most suited to the needs of the individual children and the interaction required.

Dolphins were not the only non-humans present at sessions. A stork was in permanent attendance. I often wondered if it was the same one at all sessions but probably the duty was shared. These birds' daily representative nonchalantly paced the central wooden walkway, which separated the swimming area from the holding pens, easing slowly towards the dolphins' fish box. I never saw one of them actually manage to steal a fish. They always seemed to lose their nerve at the last minute and turn tail. They would slowly walk away a few steps, their proud demeanour masking their criminal intent and then they would begin the attempted larceny all over again.

On one very hot afternoon, a disturbingly large iguana, measuring about a metre (3 to 4 feet) from head to tail, lumbered along a low hanging branch of a tree at the edge of one of the holding areas. The visitor caught the attention of both humans and dolphins alike. One of the dolphins speedily lunged at the branch knocking the iguana off balance and into the water. I couldn't see what followed but the mêlée of splashing and thrashings was a little alarming. One of the trainers informed a small group of concerned adults and children that the dolphins were just playing with it. However, his half hidden shrug and grimace conveyed a different story to the parents in the group.

Individually tailored programmes are drawn up for each child based on the information supplied by parents and specialists prior to attendance at the programme. Each child is encouraged to achieve tasks appropriate to their developmental needs and abilities. These may include, for example, saying individual words, putting words together in sentences, making certain movements, raising a hand or touching a ball. The interns make detailed notes on the session so progress can be closely monitored.

The dolphin therapists come from a range of health-care and related academic backgrounds: psychology, speech therapy, occupational therapy and neuro-science, for example. They are specifically trained in DHT by Dr Nathanson and his supervisors. One thing that really struck me about the therapists was their enthusiastic but sensitive support for the children. I don't want to make too much of a generalisation, but their very American articulations of encouragement, such as 'Good job' and 'All right', were very fitting . It seemed to me that the therapists' approach knowingly or unknowingly emulated the sensitivity and exuber-ance of the dolphins.

Dr Nathanson spent most of his time during the ses-sions observing and talking to parents, but on several occasions during my visit he went on to a station and worked directly with a child. The nature of the dol-phin interaction 'reward' ranged from stroking and touching a dolphin from the floating platform to dor-sal fin and 'belly' rides in the water. These were always very closely supervised by the therapists and trainers, and children who were more severely disabled were fully supported so they could ride too.

Families are encouraged to bring their children for a minimum of two weeks but some come for longer. The first day of the programme is used for induction and offers parents and children a chance to ask questions and familiarise themselves a little with the programme. There is no in-water activity on this day. Many families bring their other children with them and I must compliment the DHT team for their efforts to engage these children too. The cost of a two-week programme in April 1998 was $6,200. I think it is particularly noteworthy that many of the American medical insurance companies (not known for their generosity of spirit) cover the cost of DHT under certain policies.

For those of you who have only just begun breathing again after being confronted with the cost of this therapy, I should explain a little further. Dr Nathanson does not own the facility he uses and therefore has to pay rental charges for the stadium, dolphin trainers and dolphins, in addition to other running costs. Included in the therapy fees is the cost of an individually tailored video to help sustain the effect of the DHT once a child has returned home. The video contains footage of the child's dolphin therapist describing exercises and encouraging the child in particular tasks. It also shows the child interacting with the dolphin which can be used as a substitute reward for the real thing.

Given the cost and practical difficulties of this form of animal assisted therapy, one may ask why use dolphins? In answering this question, Dr Nathanson points to the comments of many parents attending the programme. Most of the children receiving DHT are involved in other animal therapy programmes too. Even though no studies have been conducted which compare the effectiveness of dolphins to other animals, parents have indicated that DHT is dramatically more effective than other more accessible animal programmes in eliciting increases in attention, language and motor skills. Dr Nathanson offers a number of possible explanations; this may be due to the novelty of dolphins, the fact that they are in the water, the ability of the dolphins to provide a wide variety of responses as rewards, other unknown causes or a combination of factors. He personally believes that dolphins have more elements about them that are pleasurable and therefore attention-holding. In my interview with him he developed this point further: 'We're interested in things that give us pleasure . . .

let's look at it through the eyes of a severely disabled population, looking at it at a very basic level. These children orient to the world as a function of their disabilities primarily through touch, actual sensations, visual sensations, auditory sensations. When you touch a dolphin it's a very pleasurable sensation. When you look at a dolphin their movement is very melodic and that melody rivets your visual attention to the animals. The sounds that the dolphins make are unique. The kids are attracted to those sounds. They like them. And of course the feel of being in the water with the animals as well combines to develop what I would call an optimum attention/pleasurable experience which is the basis of course for what we want to do because then we can get to the next step, which is the learning. Also the dolphins have this permanent smile on their face . . . that's pleasure. People smile when they see this, they're attractive, there's an aesthetic quality to them which is interesting.'

The Families

The DHT office has arrangements with two local hotels and agents for a nearby apartment complex which makes finding accommodation a painless affair. Things only begin to hurt when you check out the hotel prices and the apartment rents, even with the discounts negotiated by the DHT staff. All of the accommodation was of a high standard and situated close to the DHT office on Key Biscayne, about ten miles from the centre of Miami. It is a beautiful and safe location but this is reflected in the ever increasing hole in your pocket. When flights and two weeks' living expenses are added to the fees for therapy, it is

easy to see how the cost for a British family can reach
£10,000.

All of the families interviewed were happy for me to
use their real names. In addition to the interviews, I
contacted each of them a month or so following their
return for an update and to see how they all were.

The first people I spoke to were the Miles family
from Scotland, who were staying in the same hotel as
myself. Alan and Caroline Miles were with their two
children, Hayley aged five and a half, and her brother
Scott, who was two. Hayley had a multiple diagnosis
which included cerebral palsy and epilepsy. She was
also quadriplegic. For me, though, her overriding
feature was a beautifully engaging smile that any dolphin would be proud of.

They had heard about DHT from one of the Scottish
daily newspapers. This was their first visit to the DHT
programme and they were in the second week of their
two-week stay. Alan and Caroline arrived in Miami
expecting very little and with the attitude that if anything of significance did occur, it would be a bonus.
Caroline felt that Hayley certainly seemed to be enjoying her time with the dolphins but the weather, which
had been mixed, was an important consideration.
Hayley seemed to respond to different weather conditions, as her mum explained, 'It depends on the
weather with her. At home she won't do very much, but
if it's nice she'll do more.' The weather was certainly
changeable. The day of the interview had been glori-
ous up to the moment I had made my way to meet the
Miles family. It was at this point that the burning sun-
shine had given way to a full blown tropical storm. As
we spoke torrential rain pummelled the windows.

Hayley's therapy had concentrated on encouraging
her to bring her head up and reaching. This was no

ordinary reaching exercise. One of the dolphins would hold a soccer size ball in his mouth and rest his head on the dock. Hayley was then prompted to touch the ball. I never saw her reach for the ball but she seemed almost transfixed by the dolphin's presence – she was certainly paying attention.

I asked Hayley's parents if they had noticed any changes since she began the therapy. The answer was a definite 'yes'. Hayley now appeared to have a little more head control and had begun to vocalise sounds in a way she had not done in the past. At home she had made sounds just because she enjoyed it but they seemed quite random. Now they felt she was vocalising with more meaning. They did not claim dramatic changes had taken place but felt there were subtle differences in the way she used sounds. Alan and Caroline also explained that the improvement in both her head control and the way she held herself was of some significance. She now brought her head forward a little more. This was of some consequence for her because before the DHT she had been completely unable to control her body.

They seemed very pleased with the way therapy had gone and felt the trip had definitely been worth it, even though they considered the programme to be expensive. They had met the cost of over £8,000 out of their savings. They would have liked Hayley to have had a longer period with the programme and believed that if she could undergo this therapy for two or three months they would see a big difference. I asked them if they had any plans to return but they were unsure. They would certainly consider it but they would have to wait and see if Hayley's progress continued at home before making any decisions.

It became clear from our conversation that Hayley

loved animals. On the first occasion I mentioned the word 'animals' I saw a response and recognition in Hayley's eyes that I had not detected before. Caroline needed little prompting to offer her thoughts on the nature of the relationship between Hayley and animals; 'She loves them. She loves to touch the animals . . . wee animals seem to have a special thing between them and children, and especially children like Hayley. They're gentle. She's never been bitten. She's never been, you know, frightened by an animal. They've always been very good with her, especially the bigger animals like horses. I put her on a horse's back and she was so relaxed. It's the touch, it's the feel. She doesn't know how to touch things, so if we put her on things and she feels it again that's another sensation for her. These are all learning processes for her.'

I spoke to Caroline and Alan about a month after their return from Florida. They felt that Hayley was certainly a little more relaxed following the trip but, 'as far as her actual condition goes we haven't seen any difference'. The improvements in head control and non-random vocalising they had seen in America were sadly not sustained, but neither Caroline or Alan seemed disappointed. They still believed that DHT was effective, but not for everyone. Their final comments on dolphin therapy, in my mind, epitomised their love and commitment to Hayley; 'When you have a child like Hayley you want to try anything and everything . . . there were no guarantees [with DHT] but you've got to try it, haven't you, and if there's something else around the corner, like with a horse, or whatever, we'll try them too. We're always looking for a magic cure.'

The next family I came into contact with were the

Sutherlands, who were also from Scotland. They lived in West Lothian and were in Florida with their son Craig and daughter Laura. Craig, who was almost five years old when we met, suffers from a quite rare condition known as Soutos syndrome. General features of the syndrome include accelerated physical growth in the early years accompanied by significant delays in mental processing and motor skills. The most typical physical features are a relatively larger head in childhood with a prominent forehead and receding hairline. The eyes also appear wide-spaced.

This was Craig's first visit and when I interviewed the family they were in the final week of three weeks of therapy. It was clear that his parents, David and Ailsa, had not arrived in America with any unrealistic expectations. They, too, had first heard about DHT from a Scottish newspaper. They then spoke to the parent of a child from Aberdeen who had attended the DHT programme and this convinced them that they should give it a try. Many of the professionals who were working with Craig (speech therapist, occupational therapist and educational psychologist) were very interested in DHT and asked the Sutherlands to bring further information back with them. However, others involved in his care were less encouraging and Ailsa felt that at times they had implied that the family didn't have anything better to spend their money on.

Craig was a very friendly and engaging little boy who had obviously inherited hefty drams of his mother and father's Celtic charm. His therapy focused on working on his concentration, his general physical motor ability and in particular his hand control. In addition, he was encouraged to use words to make sentences. I watched him, one very hot afternoon, practising writing the letter 'C'. He worked extremely hard at this with support from Donny, his therapist. His reward was a well deserved full on, high speed dorsal fin ride. As he rode the waves with the dolphin he called out to his dad, 'Daddy watch me, look at me.' His father, David, excitedly commented that this was the first time that he had called out to him while in the water and it was indicative of how his confidence had grown since arriving at the programme. It was a lovely moment and one that I am sure will be cherished for some time to come.

David and Ailsa had noticed numerous changes. Craig now possessed a spontaneity which they felt was lacking before. His concentration span had definitely lengthened and he was putting sentences together instead of single words. Donny had continuously encouraged him to, 'Look at me when I'm talking to you.' As a result, it appeared that Craig's 'social eye contact' was also much improved. Ailsa felt he'd really enjoyed himself too. Overall, the Sutherlands had found the whole DHT experience very positive indeed.

David and Ailsa praised Donny for his attitude and enthusiasm. They described his hands-on encouragement and verbal rewards – 'Gimme five' etc – as significant contributing factors to Craig's progress. They were very pleased that they had come for three weeks and believed this had given Craig time to settle, to get to know Donny a little and feel comfortable with the dolphins. When I asked them if they would come again I received a most emphatic 'yes'. I enquired about Craig's relationship with other animals. Ailsa explained that they had used time spent horse riding preparing Craig for the dolphins. He needed time to develop confidence with animals. He only agreed to go on a horse for the first time a few months before the trip to Miami. He overcame his fears and now really enjoys it. They were hoping to take him on a regular basis when they returned to Scotland.

When I spoke to Ailsa back in Britain she certainly felt the progress Craig made in Florida had been sustained. Friends and family have noticed that he is saying more and his words are clearer. However, as yet, no professionals had commented on these developments. Ailsa also finds him easier to manage when he is upset because of the improvements in his ability to communicate. To a certain extent, he is now able to express his needs more effectively, which reduces levels of frustration for Craig and his family.

The Sutherlands were unable to meet the cost of the dolphin therapy trip from their own resources. Money was raised through the efforts of a fund-raising committee of friends and with the support of family. They were extremely grateful for the support they received and touched by people's generosity. They are hoping to make another trip to Florida for further dolphin therapy with the remaining money in the fund in the very near future.

My relationship with the next family I interviewed may have caused our American hosts a degree of concern. We got on extremely well but some of our verbal exchanges may have appeared less than friendly. I think I'd better explain. The Warwick family were from just north of London, only a few miles from my office. Tony and Dawn Warwick had brought their son James, who suffers from a developmental disorder, 'global delay', to the DHT programme. They were also accompanied by Dawn's mum and dad and James's two older sisters. It quickly became apparent that the male members of the family supported Arsenal football club. Unfortunately, I have been a life-long supporter of their arch North London rivals – Tottenham. Thankfully, some of the more enlightened female members of the family shared my allegiance. As a consequence, our conversations were often laced with jibes and caustic comments which, to the uniniti- ated or uninterested, may have seemed quite worrying and/or bemusing. Football allegiances aside, the Warwicks were a very warm and friendly family and it was a pleasure to make their acquaintance.

James was nine and a half years old, but mentally it was estimated that he was only three and a half. His condition profoundly affects his speech and he also suffers from asthma and colitis. He has some words but rarely uses sentences. He is a very lively boy and behaviourally can be quite challenging. This was his second two-week visit to the DHT programme. He had travelled with just his parents the previous year. The family's interest in DHT was raised when Dawn came across an article about a little boy having dolphin therapy in a magazine. She was a little stunned because the boy's condition was so similar to James's that she felt it was almost as though she had written it herself. She contacted the author of the article to find out more.

Dawn initially described to me what had happened on their first trip. During the first week Jamie had learnt to swim and thoroughly enjoyed himself and then, during the second week some significant devel- opments became apparent, '. . . There were words coming out that we hadn't noticed before and he was trying to say a lot more . . . his sentences were between three and four words . . . In the last week he said "Can I have a drink please, Mum?" – seven words, which is the longest he's ever said.'

Jamie managed to sustain this improved level of communication on his return from the first visit. Originally the staff at James's school were rather sceptical about dolphin therapy but after hearing him speak on his return, they were more accepting of the treatment. Some of the teachers were quite astonished by his progress. Dawn related to me how one of James's doctors reacted when seeing James for the first time after the Florida trip; 'Dr Hannah at Chase Farm couldn't believe the difference . . . James came in the surgery and got a puzzle out, did the puzzle, sat down and started to read a book and he (the doctor) just fell back on his chair and went "Where have you been with this child?" He couldn't believe the difference. I said dolphin therapy. He then told us, "I don't think he's got Global delay anymore. If he had Global delay he would not be able to do this".'

James's condition was now diagnosed as a 'communications disorder', but finding the appropriate treatment was still a problem.

Tony and Dawn were keen to build on the progress made in the first visit. When I met the Warwicks they were only two and a half days into their first week so it really was early days. The dolphin therapist was working with Jamie to develop his eye contact. They were already seeing improvements in this area. This work was often quite tough for James because he wanted to be in with the dolphins. James exhibits his displeasure quite readily and the therapist had to be very firm with him but he was becoming more at ease and was starting to concentrate more. James struck me as a very determined and strong willed little boy. Given the difficulties and frustrations he has to contend with due to his condition, I believe he behaves as well as he can. Dawn and Tony were full of praise for the DHT staff,

'The dolphins are really lovely but I think the way they work with the children they are, you know, special people. They've got so much patience and they really need it with James because he's so crafty. They just get their attention and they hold it . . .'

The Warwicks, like the Sutherlands, had to fund-raise to meet the cost of the trip. In twelve weeks they had collected enough money for two trips. They, too, acknowledged the generosity of friends and strangers who donated money. They would like to come again, but for longer if possible. DHT does seem to offer James hope while mainstream approaches have had only limited effect on his condition. James's parents and grandparents separately described to me how at ease he was in the company of dogs too. He shows little fear, even when meeting rottweilers, bull mastiffs and the like. He is very close to Jesse, the family dog, and he seems to converse more readily with him than people. Dawn believes there is a very strong bond between them, and animals in general definitely do help James.

I spoke to Dawn and Tony about five weeks after their return and James seemed to be doing well. At his yearly review at school staff remarked that, following his time with the dolphins, they had noticed an improvement in his concentration and that he seemed to be calmer. The family felt he was trying to talk a lot more and was using fewer signs. His eye contact had steadily improved during the two weeks in Florida and this had continued since their return. Dawn also believed that he was keener to work than he was before the trips but was still very crafty when it suited him. I think it would be a shame if he lost this quality completely and I'm pretty sure his family feel much the same.

*

My final interview was with Geraldine Graham, a nurse and mother from Northern Ireland. She was a single parent who had brought her four-and-a-half-year-old daughter Natasha across the Atlantic with her cousin. Natasha has multiple disabilities and her diagnoses include quadriplegic cerebral palsy, epilepsy, brain damage, scoliosis, poor vision and bilateral dislocated hips.

Geraldine was feeding Natasha when I interviewed her and she explained to me that just prior to becoming pregnant with Natasha, she had been caring for a child with cerebral palsy. She felt she may have been a little more prepared than most to care for a child with the condition but nothing can really prepare you when it's your own child. She had raised Natasha alone since birth with little respite. Her family have been supportive but she had to move away from her hometown to Belfast in order to be closer to treatment facilities. It quickly became apparent when talking to Geraldine that she is not bitter about her situation at all and, as she related, she sees no point in complaining. For the first time in one of my family interviews in Florida, I slipped into work mode. I explained to Geraldine that I was a psychologist by trade and that I strongly recommended she took time out to complain occasionally because we are all entitled to gripe now and then and she had more to contend with than most. Her response didn't surprise me: 'I wouldn't have time to complain even if I wanted to.'

She, too, had heard about DHT in the press. She thought it was well worth a try and she was now in the third week of a four-week stay. Geraldine had certainly noticed changes as a result of Natasha's dolphin therapy. She felt she had a 'special' eye contact which she hadn't got before. In addition to this improvement

in her 'eye control and contact', it seemed that her
body movements had developed too. She was now
reaching out for things. Geraldine also believed that
her concentration span was a little bit longer. Similarly,
her sleeping patterns had got better and her seizures
had reduced. Overall, the most significant development was with her eye contact, as Geraldine explained,
'She used to look with her right but her left eye was
always dropped. It never focused on anything but the
two will actually look at you now and if you ask her to
look she will look at you straight in the eyes . . .'

Geraldine was obviously very pleased with Natasha's
progress and in good spirits. However, her demeanour changed somewhat when I asked her about
the attitude of her fellow medical professionals towards dolphin therapy. She made it abundantly clear
that they had not only disagreed with dolphin therapy
but were against other complementary approaches she
had tried. She had taken Natasha for aromatherapy,
oxygen therapy and to see a chiropractor. Had they
helped? Geraldine was in no doubt, 'Definitely. The
oxygen therapy has, it helped her to chew which she
couldn't do before. It was all puréed foods but now she
can chew. Her dribbling has stopped. She used to drib-
ble constantly and she would use up to a hundred bibs
a day. She doesn't use them any more. Her sleeping
pattern, it helped with that as well . . . Her eyes became
very clear. They used to be very grey and bloodshot
and her circulation was very poor as well and it has
improved dramatically.' She also felt that the aromatherapy had very much helped Natasha to relax.

The treatment Natasha received from the chiropractor had a most profound effect on her scoliosis – a
curving or crookedness in the spine. Geraldine
explained that this had been placing pressure on her

rib cage which, if it had been left, could have resulted in her lungs being crushed. Geraldine described to me what happened when she took Natasha to see her consultant following the treatment with the chiropractor, 'We went in March for an appointment and he was checking her back and he said, "What's gone on here? I've seen a child get a little better or get worse or stay the same but never one who had recovered almost completely from scoliosis".' Geraldine explained that the spinal problem had disappeared but some muscular problems remained. The consultant was dumfounded, as was Natasha's physiotherapist. She had told Geraldine that there was no point in trying these therapies because Natasha was never going to get better. Even though Geraldine had already proved a point to these care professionals, they were just as negative about dolphin therapy and considered it a waste of money.

I asked about Natasha's relationship with other animals and Geraldine described the very special support her sister's dogs give her. 'My sister's got two King Charles spaniels and they seem to sense that there's something wrong with Natasha because they don't bother with the other grandchildren, just Natasha, and they'll come over to her and lick her. They'll lie up beside her and if she's sick nobody's allowed to touch her, especially with Monty. He lies beside her, licks her face and will just cuddle into her and in bed at nights she'll go up, the dog will go up to bed with her. I'll lie Natasha down and he lies beside her until she goes to sleep and then he comes downstairs again. Oh they're unbelievable. They just sense that there's something special with Natasha.'

Geraldine had to raise funds for the trip herself and did this by organising charity evenings and dance nights. She also ran a marathon with friends in a relay. Unfortunately, her fund raising was initially hampered due to the sectarian divide in Northern Ireland. One side of the community had been reluctant to contribute to the needs of a child from the other. Through her determination she overcame these difficulties and the funds were eventually raised. Hopefully, with the new found spirit of reconciliation in the province, such pleas for support in the future may help bring the two communities together.

The only criticism Geraldine expressed about the DHT programme was it's short duration, but she was full of praise for the staff. There is also no question in her mind that, funds permitting, she will return for more dolphin therapy. She intends to come every year for the next five or until Natasha begins to talk. I have no doubt that if this objective is achievable, Geraldine Graham will ensure it comes to fruition. I wish her well.

At no point did any of the families I interviewed complain about their situations and they all generally seemed optimistic about the future. They certainly earned my respect and admiration. It is a pity that money is not more freely available to fund visits to the programme but it is hoped that attending for therapy may be less expensive when Dr Nathanson can build his own facility. A permanent base for the programme, with research facilities and accommodation for visiting families is his ultimate objective. At the time of my return to England, the DHT programme was about to be moved to Cancun in Mexico because the Miami Seaquarium was establishing an open swimming programme where members of the public would be able to pay to swim with the dolphins. As a result, no time would be available for the DHT programme. This change of venue is likely to lengthen the waiting time for DHT, which at the time of writing was already about a year.

To my mind, Dr Nathanson's dolphin human therapy programme is the world leader in this field, but similar projects have been established in other parts of the world. Over the next few pages I shall briefly describe a selection of these projects before addressing some of the thorny ethical issues raised by dolphin therapy.

Other Dolphin Therapy Programmes

*Supportive Experience with the Aid of
Dolphins – Israel*

This therapeutic swimming programme is based at Dolphin Reef in Eilat, Israel, on the shores of the Red Sea, where, over the past few years, a group of bottlenose dolphins have made their home and have successfully reproduced and raised their young. Some of the dolphins venture into the open sea to hunt while others are fed by the trainers. The philosophy of Dolphin Reef is to enable the dolphins to go out into the open sea and return if they wish.

'Supportive Experience with the Aid of Dolphins' began in 1991 and is just one of several dolphin projects underway at Dolphin Reef. Participants in the programme have included children with learning disabilities, problems with concentration or communication, hyperactive children, those suffering from post-traumatic stress and those with mental health problems, including depression and anorexia. Children with Down's syndrome, autism and those who have been victims of sexual abuse, as well as those who are deaf and blind have also taken part in the programme. This facility differs from DHT in that it only takes children who are over seven years old.

Space is very limited and therefore only a few children can be accepted for the year-long programme. At most, three children a day attend, in order to allow for the maximum amount of attention to be given to each child. The programme allows children to take part in a series of therapeutic encounters with the dolphins. Each series is made up of seven courses and each one lasts four days. Daily encounters are held lasting up to an hour in duration and are divided into two parts.

Part 1: Participation in Training

During this activity the child is invited to participate in a training session under the guidance of a dolphin trainer. The child is encouraged to approach the dolphins gradually and to become involved to the point where he or she eventually feeds and 'signs' to the dolphins. The activity is designed to increase levels of concentration and enthusiasm, and to give the child goals to achieve involving the dolphins which stimulate the thought processes and the will to co-operate. It is hoped that working with the child in this environment enables the therapeutic objectives to be addressed under pleasant and enjoyable conditions and not as a compulsory lesson. As a result, it is hoped that the level of stimulation and motivation on the part of the child is higher and affords a positive way to establish contact with him or her.

Part 2: Swimming with the Dolphins

In this part of the activity the child and his trainer swim or float together amongst the dolphins. The dolphins are never fed during this swim time so these meetings are spontaneous and unconditional. The contact between the dolphins and the child is therefore dependent on the free will of the dolphins. The philosophy underpinning the programme is that the dolphins' friendliness towards man, the pleasant touches of their noses, the range of sounds they make and the eternal smile on their faces all make the meeting exciting and stimulating for the child. The child also realises that his disabilities don't repel the dolphins as they may humans, but have the opposite effect – the dolphins recognise the unfamiliar and behave with intense interest and gentleness. Unlike the meeting during the training session, which primarily involves the thought process, this meeting in the water hopefully touches the heart and focuses on the emotional aspect.

The two types of meeting are designed to complement one another. It is hoped that in addition to raising the level of the child's thinking, they also develop self-confidence, self-worth, plus an improvement in general communications and the ability to express emotions.

The therapy team is made up of the dolphin trainers and a resident psychologist. The trainer sets goals for the child in the identified areas of difficulty. Targets which require particular emphasis are decided upon, in advance, together with the parent/psychologist/therapist who knows the child and under the supervision of the resident programme psychologist.

Island Dolphin Care – Florida, USA

This programme, located on Key Largo, Florida, in a deep-water lagoon, has been developed by Deena Hoagland to help children and their families who are living with a range of developmental, physical and/or emotional difficulties. Deena, a clinical social worker with a background in psychology, created the dolphin assisted therapy programme after witnessing the recovery her son, Joe, made when he began swimming with dolphins at the age of three. Deena and her husband brought him to Dolphin Plus to see if the dolphins could motivate him to use the left side of his body, which had been weakened by a stroke during open heart surgery. Joe had not responded well to conventional physical, occupational and speech therapies. At Dolphin Plus, he began to practise his physical and occupational therapies in a natural, non-threatening and fun setting. He took pleasure and satisfaction in completing his exercises, assisted by the support of the dolphins. Joe responded extremely well to this therapy, increasing both muscle tone and flexibility, as well as developing a more positive level of self esteem. With the support of the dolphins, especially his particular friend, Fonzie, he was able to normalise his life.

Deena believed that if dolphins could help Joe feel better about himself and motivate him to try new tasks, then the dolphins might help others. Over the years Deena has worked with many different children with a range of educational, emotional and physical needs. She believes that the dolphins become catalysts that provide support for these children. Metaphors are drawn from the dolphin–child interactions during sessions which follow each swim. Positive behaviours are reinforced by these interactions, with the aim that the results are carried over into school, the home and the community. Deena and her colleague, Dr William Shannon, have worked with pre-school children with learning disabilities, and their parents. In addition to enabling the children to be more expressive, this pro-gramme has helped to develop a closer bond between parents and their children. Deena and William also work with groups of adolescents with issues of abuse, problems at home or school, various emotional diffi-culties and/or low self-esteem.

The therapy programmes are set up in five-day seg-ments running from Monday to Friday. Participants may attend for one or more weeks. The five-day pro-gramme consists of five platform/in-water sessions with the dolphins lasting between 20 and 30 minutes each, optional family swimming with a group of dolphins and four individual/family psycho-educational class-room therapy sessions lasting up to 50 minutes. The programme is designed to assist children and adoles-cents who have chronic medical, mental or physical disabilities; those who are dealing with emotional problems; children with attention deficit hyperactivity disorders (ADHD) and those who need to strengthen their sense of self-worth and self-esteem. The family is also an important focus of the programme.

Platform/In-Water Sessions

These sessions are based on a play therapy framework. Deena and her colleagues firmly believe that children learn the most through personal attention and play. During these sessions the participant will work one-to-one with a dolphin. The dolphin handler, the therapist and the recipient will be placed on a platform during the session. Activities from the platform are based on the needs and abilities of the participant. They range from hands-on games with the dolphin to in-water games with the therapist and the dolphin. Activities learned in the classroom are reinforced on the platform and in the water. These experiences can be so rewarding that they stimulate the child to attempt things they might otherwise hesitate to try. The relationship developed throughout the week with the dolphins should increase levels of self-esteem, attention, commitment, trust of others, and provide motivation to continue to attempt new tasks.

Unstructured Swim With the Dolphins

Participants capable of swimming on their own can take part in swimming with a group of Atlantic bottlenose dolphins. Participants under the age of 18 are accompanied in the water by an adult, parent or guardian. This is an opportunity to view the dolphins in their world. For about half an hour swimmers are able to hear the dolphins vocalising their clicks, whistles and grunts as they splash and play around the participants. Reports from the programme indicate that those taking part in this swim often feel relaxed, refreshed and invigorated. The unstructured swim is viewed as a recreational therapy where the dolphins can act as catalysts that provide a unique stimulus which may enhance trust and increase the parent-child bond.

The Classroom

Each participant spends up to 50 minutes a day for four days with a therapist in a classroom setting at Dolphin Plus. Here the individualised programme for each participant is delivered. The classroom activities are based on each child's individual needs. Activities that are educational/recreational are presented individually or in a small group. For most children, lessons follow a curriculum emphasising a marine and dolphin theme. A dolphin and marine pictorial environment has been created in the classroom to simulate the platform experience and to portray in pictures the close relationship that occurs between a participant and his or her dolphin friend. Ocean and dolphin sounds play softly in the background to produce a relaxing atmosphere, while the therapist and the participant explore new ideas in counting fish, preparing pretend play food buckets for the dolphins, assembling dolphin, fish and starfish puzzles, playing games involving co-ordination of movement, creating art projects and working on the computer. All of these activities are designed to reinforce the platform and swim experiences.

For those children and adolescents suffering from abuse, emotional difficulties, attention deficit hyperactivity disorders (ADHD), or poor self-esteem, lessons follow a curriculum emphasising: *i)* education about their individual problems; *ii)* knowledge about how dolphins can help an individual to express his or her problems more freely; and *iii)* strategies to assist in overcoming their difficulties. Before being accepted on to the programme, detailed medical and psychological reports are required.

Conny-Land, Switzerland

This is a very new programme located in Lipperswil, Switzerland, close to the border with Germany. The programme, situated in an enclosed indoor facility, commenced in January 1998 and is supervised by Elisabeth Palfalvi, a senior physiotherapist. The main dolphin pool is circular, 16 metres (53 feet) in diameter and 6 metres (20 feet) deep. Elisabeth works in a similar way to David Nathanson, using the dolphins as both a stimulus and a reward to increase the attention span and develop the motor and speech skills of children with a range of disabilities and special needs. At present, she conducts a one-hour session twice a week but she is hoping to increase this to five days in the very near future due to an ever increasing demand. Elisabeth is keen to point out that she does not feed the dolphins during their time with the children. They join the therapy sessions without any inducement. Elisabeth is supported in her work by two other therapists and four dolphin trainers. The supporting therapists have skills in physiotherapy, massage and therapeutic breathing techniques.

The children interact with five dolphins from a platform and in the water. Parents are invited to join their children on the platform. This enables them to observe Elisabeth's methods and learn new approaches to interacting with their children. Sessions last approximately an hour but they can vary between 50 and 90 minutes depending on the individual needs of the child. Videos are made of each session and copies are given to parents. This allows everyone involved an opportunity to review exactly what has happened during the dolphin encounter.

Parents are initially asked to book a minimum of

three sessions over a two-week period to allow their child to become accustomed to the environment. More than three sessions can be booked, but if Elisabeth feels this is too much too soon for a child she will review the situation with the parents. Before a child can be accepted on to the programme, Elisabeth makes a home visit to see the family situation and discuss former treatments and therapies. This visit also offers parents an opportunity to ask any questions they may have before the programme commences.

It's very early days for this programme but Elisabeth proudly informed me that every family who have attended for therapy have booked further sessions.

Ethical Considerations

If, at this point, it appears that I'm getting on my high horse a little, please bear with me. In the introductory paragraph of this chapter I briefly described my feelings regarding the well-being of both the dolphins and humans involved in therapeutic programmes. I intend now to explore the respective species' plight in turn.

Firstly the dolphins. For many, the primary ethical consideration relates to the issue of captivity. The confinement of dolphins has given rise to very heated debate and has culminated in angry demonstrations outside dolphinariums. Many animal welfare activists have argued that in no circumstances at all is it ethical to restrict dolphins' freedom, while others have suggested that the nature of the captive environment must be carefully considered before a judgment can be made. Dr David Nathanson is a firm believer that there is good and bad captivity. He points to studies that have indicated that dolphins who live and breed in well planned captive environments which offer good quality food and ongoing veterinary care are generally in better health and have longer life spans than those that live in the wild. He also cites post-mortem reports that have identified more stomach ulcers in wild dolphins than in dolphins who have lived and died in good captivity as an indication that the latter suffer lower levels of stress.

Those against any form of dolphin confinement consider the notion of good captivity a contradiction in terms and point to rival studies which suggest that dolphins are less happy and less healthy in these 'man-made prisons'. This debate is sure to continue and I am a little torn on the subject. In an ideal world I would like to see all living creatures free and unconfined. However, how would this work in relation to cats and dogs, for example, who also live in confining human environments and are perceived, in the main, as being owned by their human companions?

Some animal welfare activists are willing to accept the keeping of a few hundred dolphins in good captivity because such human contact may have benefits for the thousands of dolphins in the wild, the two main ones being: *i)* awareness of marine conservation issues can be raised by educational encounters with dolphins in such facilities; and *ii)* controlled research studies can be conducted to address issues of dolphin health and welfare. Those individuals and groups who are completely against any form of captivity reject the first of these by arguing that awareness can just as easily be raised through the media and educational resources. This may be so, but it has also been suggested that the impact of first-hand experience is so powerful that those witnessing dolphins for themselves need no persuasion on the importance of marine conservation and welfare issues. I don't believe it is either practical or desirable for large numbers of people to have encounters in the wild. If this was attempted, it would necessitate many boats invading dolphin environments which could be detrimental to the dolphins because of the resultant increased noise and petrochemical pollution.

Some observers believe the second benefit is even more difficult to discount. How can controlled research studies be conducted in the wild? The three dolphins we met at Flamingo Land, Lotty, Betty and Sharky, were involved in a study undertaken to help prevent the deaths of many thousands of wild dolphins and porpoises who are caught and drowned in fishing nets. The research, under the directorship of Peter Bloom who ran the facility, involved Cambridge, Aberdeen and Loughborough universities. The captive dolphins' echo location systems were used to develop sonar reflectors to attach to fishing nets so dolphins and porpoises could detect them and take avoiding action. Short of banning nets, which some may feel is desirable, the lives of wild dolphins could not have been saved without controlled studies with captive dolphins.

Some facilities are run on the basis of semi-captivity. Dolphin Reef in Eilat, for example, attempts to offer the dolphins a choice. However, for many existing facilities this would be impossible due to the physical environment. Another possible stumbling block concerns the fact that considerable investment is required to develop a facility and if a semi-captive approach is adopted financing problems could arise. If the dolphins are allowed to exercise their choice and leave, or their attendance is irregular, programmes may be halted or interrupted and money inevitably lost. As a result, future investors may be more difficult to find.

However, it would seem that the Dolphin Reef facility works well and, where possible, the system they have adopted may be the most desirable compromise.

Moving on now to the human species. I don't know if it's irresponsibility or over enthusiasm, but a number of media reports in recent years regarding dolphin therapy have caused me a good deal of concern. A couple of examples which immediately come to mind relate to the supposed curative effects of swimming with dolphins on people suffering from the eating disorder anorexia. I don't need to consult my records to recount these tales because the headlines of these reports are indelibly imprinted on my memory. The first read, 'Swimming with a dolphin cured my anorexia', and the second was one of a series of articles entitled 'My favourite cure', beneath which sat, in this particular instalment, a description of how swimming with a dolphin was that person's favourite cure for anorexia. I am not denying that some therapeutic benefit may have been derived from their experiences, but a cure for this potentially life threatening condition? I don't think so. In fact when one reads between the lines of these reports, it soon becomes apparent that these experiences were far from curative. This is no surprise given the difficulties associated with treating anorexia. Such headlines often serve to raise the expectations of the sick and their families. As a consequence, painful disappointment can compound already existing problems.

Similarly, I must challenge the concrete assertion made by certain commentators, that dolphins always offer 'unconditional love' to those who are in particular need of such attention. I am not denying that this can happen, but Pat Morell, the therapist at Amble, who has taken many people suffering from mental distress to swim with wild dolphins, is unequivocal in her view that incidents of dolphins directing unconditional love and attention to those in great need of support are a more sporadic occurrence. To raise the expectations of depressed individuals with low self-esteem and self-worth by suggesting that they will receive unconditional affection and acceptance from a dolphin, and to then expose them to what appears to be rejection from a creature that is supposed to love one and all, is irresponsible to say the least and may result in tragic consequences.

Finally, as I commute between my ethical high horse and soapbox, a question which has repeatedly come to mind in my exploration of dolphin therapy – how are the dolphins effected by their involvement in human healing? It would appear that no one really knows. Close monitoring of dolphins' health certainly gives some indication of their physical well-being. However, I believe the effect on the personal and social development of dolphins to be a crucial consideration and one which should be uppermost in our minds when we ask them to leave their more natural environments and take on the role of therapist.

Even though this chapter is at an end, I can assure you this is only the beginning as far as animal assisted therapy is concerned. Dolphins may conjure up wonderfully romantic images, but meeting them may necessitate travel and overcoming a range of practical and financial obstacles. But not to worry, because other healing animals are close to hand. In the next chapter we meet man's best friend. However, before I move on to the canine world, I would like to conclude my dolphin thoughts by reiterating my respect and admiration for these enchanting creatures, the therapists and, of course, the parents and children.

CHAPTER 5

MY BEST
FRIENDS – DOGS

The dog is by far the most common supplier of animal assisted therapy. Their contribution is staggering and includes guide dogs for people who are blind or visually impaired, hearing dogs for deaf people, dogs for people with physical disabilities, visiting PAT (pets as therapy) dogs and even dogs who are capable of predicting epileptic seizures. These canine carers are collectively known as service or assistance dogs and the modern origins of their contribution to human well-being can be traced back to the First World War. The two world wars saw Red Cross dogs bring help to many wounded soldiers. In addition to the enormous number of human casualties of these conflicts, over 7,000 British, American and German dogs were killed in action. It was in these sad and horrific circumstances that humans really began to recognise the enormous potential of dogs to support them in the most adverse situations. I will begin my exploration of assistance dog groups with one of the oldest and best known organisations:

Guide Dogs for the Blind (Seeing Eye Dogs)

I feel a special affinity for Guide Dogs for the Blind because some of my earliest memories suggest that I spent most of my childhood in pursuit of used, and often rather smelly, foil milk bottle tops to donate to the organisation to raise funds. These recollections were, to some extent, substantiated when my research revealed that in the 1960s over 20 million bottle tops were required to fund the training of one dog! As a child I thought they were melted down to make money, a feat I attempted on a couple of occasions with little success, but I digress.

The Germans were the first to train dogs in an organised way to lead blind people in the 1920s. This was in response to the needs of thousands of their soldiers blinded in the First World War. In 1927 this work came to the notice of an American woman, one Mrs Dorothy Harrison Eustis, who herself was training police and army dogs in Switzerland. Her kennel manager then spent the next year or so learning the German methods and training a guide dog for a fellow American, Morris Frank. He travelled to Switzerland for training and on his return became the first American with a guide dog and this was how Seeing Eye Dogs was born. In 1930 two British women, Muriel Cooke and Rosamund Bond, contacted Dorothy Harrison Eustis and expressed a strong interest in her work with Seeing Eye Dogs. Dorothy very kindly sent a trainer from Switzerland to England and a year later the first four British guide dogs were trained and ready for work. In 1934 the Guide Dogs for the Blind Association was established in Britain. The organisation has since grown into one of Britain's best known charities. In 1996 there were over 4,000 working dogs and 720 in training. They have an extensive breeding programme (the largest working dog breeding programme in the world) and place puppies at six weeks old in a family home which provides a firm foundation for later training. The puppies are brought up gradually to deal with increasingly noisy and busy environments.

Any visually impaired person over 16 can apply for training at a residential centre which lasts three to four weeks. Great care is taken to match the right dog for each individual. The overall objective of the training is to teach virtually impaired people how to use their dogs effectively. The dogs' training has five primary objectives: *i)* to walk in a straight line in the centre of the pavement unless there is an obstacle; *ii)* not to turn corners unless told to do so; *iii)* to stop at kerbs and wait for the command to cross the road; *iv)* to judge height and width so that its owner does not bump his head or shoulder; and *v)* how to deal with traffic. Ninety-five per cent of the dogs trained are Labradors, golden retrievers or Labrador/golden retriever crosses. The remaining five per cent are mainly German shepherds and Border collies.

As well as providing blind people with practical support, an American study conducted in the early 1980s suggested that the companionship of a guide dog provided additional benefits. Forty-four guide dog owners were surveyed and the findings indicated that the dogs not only offered their human companions greater mobility but also gave them an increased ability to cope with their blindness. Other positive aspects of guide dog ownership were found to include: acceptance of life and risk taking, expression of feelings, assertiveness, personal achievement, orientation to the present, relaxation, improved body image, security, self-control, self-awareness and opportunities for social outlets. Guide dogs have also enabled blind people to undertake a wide range of employment. This has certainly been the case for Alex, an administrative supervisor in an extremely busy London office, with whom I met to record his experiences as a recent recipient of a guide dog.

When I met Alex he was partially sighted with severe tunnel vision, total night blindness, no or very little colour vision and cataracts. Unfortunately, he had recently been advised that he was very likely to lose his sight completely within the next two years. He decided to apply for a guide dog while partially sighted because he felt that the little vision he did have might enable him to become more readily accustomed to this new means of support. While we were together, Angus, his two-and-half-year-old yellow Labrador was at play. It was hard to believe that such a lively, sociable dog could have the discipline to perform a guiding role but I was assured by Alex that he was certainly up to the job. They had been together for eight months and it was clear that a very strong bond had been formed.

Alex underwent three weeks' intensive instruction at a residential training centre to prepare for his new companion. Training took place six days a week, plus three evenings. This was followed by a further three weeks' training at home. Additional support was available on request after this period. The experience had proved quite challenging for Alex but he felt he had coped well. However, he was aware that others on his course had found it far more arduous. A number of areas were covered in the training, which included dog behaviour, how to clear up after your dog, canine first aid and how to administer or force-feed medicines, plus how to handle your dog as a guide dog and as a pet. All trainees worked through at least two practical exercises each day. I was particularly interested in the differences between handling Angus in his working and pet roles. Alex explained that two totally distinct sets of commands were used which made it clear to the dog what was expected of him.

As I mentioned earlier, most guide dogs are Labradors and retrievers but occasionally German shepherds and collies are trained. Future recipients select their preferred breed and Guide Dogs for the Blind do their best to accommodate their wishes. Alex had chosen a Labrador and it was abundantly clear that he was more than satisfied with the breed's representative he had been allocated. Angus was a most appealing dog, both at play with his lolloping stride and boundless energy and at work as a skilled and committed guide dog.

Alex spoke frankly about the difficulties he and others had experienced during the training period. He feels that undergoing training away from one's own living environment is a major stumbling block and he believes that the training would more effectively meet individual needs if it took place in one's home and local area. This approach would enable recipients and dogs together to begin to become accustomed to their real living environment. It must be said that Guide Dogs for the Blind Association would welcome such a shift in the training process but are unable to implement such a change due to lack of resources.

There is clearly a period of adjustment following the formal training. The new owner and the dog have to find their feet (paws) together. Alex described how he very quickly found himself bending some of the rigidly enforced handling rules that had been imparted during his training. When he mentioned his increasingly flexible approach at his progress review, he was a little surprised at the response he received. His trainers informed him that this was absolutely fine and only to be expected. To be honest, Alex was a little put out that this developmental process was not mentioned during the course. If this had been brought to the trainees' attention, they would have been aware of possible changes and not unduly concerned by the unexpected.

The progress of an owner and guide dog's relationship is reviewed annually, but if problems occur the training team respond quickly to requests for guidance. Alex certainly experienced some teething problems with Angus but these have been heavily outweighed by the advantages of having his support. Angus has given Alex total independence. He has now been able to overcome the difficulties he experienced with his night blindness, and as a result, he no longer needs to leave work early to ensure he gets home safely. He also feels that this rather special dog has both helped him come to terms with his loss of sight and conquer some of the pitfalls facing blind people. I asked Alex to elaborate on the potential pitfalls and he responded with a most succinct reply, 'the general public'. He explained that the way many people have responded to him and Alex in the street was at best insensitive and at worst positively dangerous. Distracting a dog when he is guiding his blind owner across the road could have fatal consequences. Such has been Alex's level of concern with these issues, that he made a short television programme on the subject for Carlton, which was shown in most parts of the British Isles. He made a polite but heartfelt plea to the public to think a little more about their behaviour when they come across a blind person with a guide dog in the street. He also detailed some dos and don'ts which are included in the Practical Guidelines section at the end of this book (see page 280).

The working life of a guide dog is usually seven to eight years, but in London it's more like five. Alex hopes to keep Angus until he retires, even though he has one minor deficiency – he hates water. Going out in wet weather has proven to be a bit of a problem. Angus has just about coped with it raining on his back but he has led Alex on a merry dance zigzagging around puddles. It struck me on hearing this that going for a walk in the rain with Angus after a few drinks could be quite precarious. I put this to Alex and he agreed, adding that he would also find it very difficult to deny the precedent setting offence of being drunk in charge of a guide dog.

My final question to Alex concerned his thoughts on the future, both in terms of his complete loss of sight, and his relationship with Angus. I think his own words best describe just how important he views his dog's support over the next few years: '. . . I mean there is a future until he retires, which is great as far as I'm concerned. I love him to bits and he is a good pet as well as a working dog. I'm so glad I made the move [to get a guide dog] and I'd recommend anyone to have a guide dog if they think it's appropriate and even if they don't think they could cope with it, they should still find out about it.'

Guide dogs are now an accepted feature of day-to-day life in Britain, but this has not always been the case. The general public of the 1930s was very much against making dogs work and dogs were not used by the police or the armed forces in the way they are today. The pioneering guide dog trainers were verbally abused in the street and people actually attempted to physically stop them from training the dogs. To many observers the work of the trainers was seen as cruel and useless. However, as the public realised the positive effects the dogs had on the lives of blind people, attitudes began to change. Nevertheless, I think Alex would argue that there is still a little bit of work left to do.

Hearing Dogs for Deaf People

Hearing Dogs for Deaf People was established in 1982.

The aim of the organisation is to train dogs to assist severely or profoundly deaf people. The dogs are trained to alert deaf people by touch, using a paw to gain attention and lead them to the source of the sound. These sounds can include alarm clocks, doorbells, cooker timers, telephones, smoke alarms and baby alarms. For sounds that signal danger, the dog will touch and then lie down to indicate the emergency.

Potential owners are carefully assessed for suitability. They must have the desire to look after a dog. Highly skilled trainers select dogs which exhibit a high level of intelligence, a friendly nature, keen responses to sound and a willingness to please. Dogs come from a variety of sources. They may be unwanted pets offered by breeders or come from rescue centres. The dogs are then trained in a homelike environment to recognise the sounds chosen by its new owner. The training is rigorous but achieved through praise and reward, and takes about four months. One week before the training is completed, the dog's owner is invited to spend the remaining time in a guest flat at the training centre in Lewknor, Oxford. This enables the owner to familiarise themselves with their new hearing companion. Further training and acclimatisation takes place in the owner's home.

The recipients of hearing dogs have reported numerous benefits, in addition to the constant and devoted companionship offered by a dog, including feelings of greater confidence, self-esteem and well-being, which have led to reductions in stress and fewer feelings of anxiety. Having a dog has also encouraged recipients to go out more often, which has resulted in them meeting new people and taking regular exercise. The dogs also provide new interest and responsibility

– a purpose for living – and have led to recipients being more readily accepted into the hearing community, which in turn has led to an increased frequency in visitors and phone calls.

These reports are supported by an in-depth study conducted over a two and a half year period. Claire Guest, operations director for Hearing Dogs for Deaf People, sent a range of health and psychological test questionnaires to 50 recipients of hearing dogs before and after the acquisition of the dog. Her findings suggest significant improvements in levels of depression, tension, social functioning, feelings of aggression, fatigue and sleeping, as a result of acquiring a hearing dog.

It would appear, from my contact with Hearing Dogs for Deaf People, that a number of the dogs they have trained have gone on to exhibit abilities way beyond the basic requirements. Two accounts from their records particularly caught my eye.

'Lady was the first hearing dog to be trained. On a very cold day her elderly owner slipped on some ice and although she managed to get back into her house and sit down, she didn't change out of her wet clothes and went into shock. Lady jumped on her lap and kept licking her face to keep her warm. Once her owner realised what was happening, she got up and put the heater on and called for help. On another cold and blustery day, Lady and her owner were going for their regular walk but as they reached a tree lined avenue, Lady stopped and refused to go down the lane so her owner abandoned their usual route took an alternative path instead. The next morning when they returned to the lane they found a tree had been blown down in the wind and was completely blocking the road.'

'Max lives with his deaf owner Shirley Anne in

Scotland. One day he alerted Shirley Anne to the sound of a fire alarm. She searched the house but no fire could be found and no alarm was sounding so Shirley Anne and her mother ignored him. He did it again and again and then led Shirley Anne's mother to the back door. When she went out into the garden, she heard next door's smoke alarm. It turned out that their neighbours had left a frying pan on unattended and it had caught fire. The fire brigade was called and they managed to save two dogs that were locked in the neighbour's kitchen. Max won a BBC 'Wag Award' for alerting his owners to the fire.'

Dogs for the Disabled

Dogs for the Disabled was established by Frances Hay in 1986. She was a dog lover who had become disabled through bone cancer. She had found that her own dog, Kim, could be trained to carry out tasks which she found increasingly difficult to do. Following Frances's death, friends and family continued her work, realising the value and happiness brought by a dog to disabled people. Current recipients of these assistance dogs are aged 18 to 84 years of age, with disabilities as diverse as cerebral palsy, polio, paraplegia and quadriplegia arising from car accidents or industrial injury, multiple sclerosis muscular dystrophy, spina bifida, arthritis and thalidomide disability.

Dogs for the Disabled provides trained dogs for people whose disability restricts them both from performing everyday tasks and leading lives as active as they would like, provided they are able to take good care of the dog according to an agreed code. The dogs are trained to do tasks that are often either very difficult or impossible for disabled people to do and they therefore enable their owners to retain a level of independence. These tasks include: retrieving a range of items – cordless phones, post and newspapers from the letterbox – and picking up dropped articles such as keys and pens; opening and closing doors; activating lights and alarms; pulling laundry out of the washing machine and fetching groceries in a shopping basket. The dogs are also trained with a special harness which enables a person with balance difficulties to walk without a frame or a stick.

Dogs are assessed and selected from many sources and include rescue dogs. All dogs are trained in basic obedience, social awareness and hygiene. This is followed by specialist training to carry out specific tasks required by their future owner. The dog and disabled recipient then spend up to four weeks together at a training centre under the guidance of an instructor. Once a dog is qualified and the recipient ready, the responsibility for the dog passes to the recipient. Instructors continue to offer support throughout the dog's working life.

June McNicholas from the University of Warwick and Dick Lane, Veterinary Surgeon for Dogs for the Disabled, surveyed 95 per cent of the recipients of assistance dogs trained by Dogs for the Disabled and found that the dogs, in addition to their invaluable support in overcoming aspects of people's physical disability, also contributed significantly to their owner's social, psychological and physical well being. The benefits included:

Improved social contact – three quarters of recipients related that they had made new friends since they received their dog. Over 90 per cent reported positive casual conversations with people who stop to talk to them when out with the dog. For one third of recipients, an overall improved social life was felt to be a significant result of having their dog.

Health benefits – more than half the recipients in the survey felt they were less worried about their health than before they had their dog; nearly 70 per cent believed they relaxed more. Just under half of the people surveyed reported that they felt their physical health had actually improved since getting the dog.

A valued relationship – more than two thirds of recipients valued their dogs as much for their friendship or companionship as they did for what they offered as working dogs. More than 70 per cent felt their dog was one of the closest relationships they had currently, while almost all of those surveyed (93 per cent) regarded their dog as a valued member of the family.

A source of support and comfort – almost 60 per cent of recipients talked to their dogs about their problems as a means of easing worry; 70 per cent frequently turned to their dog for comfort when feeling sad or upset.

These findings clearly suggest that assistance dogs significantly enhance recipients' quality of life. Two brief accounts from the records of Dogs for the Disabled clearly illustrate the life-changing effects acquiring one of these dogs can have. 'Debbie was a county standard badminton player until a block of concrete fell on her, damaging her spine. Unable to walk without crutches, her injuries prevented her from enjoying many of the normal pleasures of the teenage years. Debbie became lonely and depressed and her future seemed bleak until she acquired Elton, her assistance dog. Through Elton, Debbie gained confidence and independence. She learnt to drive and now has a full-time job as a PR officer. She has since married and now lives with her husband and, of course, Elton.'

'Anne was virtually housebound through her disabilities until Shep was trained for her. Since then she has completed a degree in computer design and chemistry, and has returned to teaching music at her local school. Anne has also been able to take part again in many of the hobbies she once enjoyed and, last year, demonstrated driving her pony and trap at the Royal Cornwall Show. As Anne put it, Shep made her take up life again.'

PAT – Pets as Therapy Dogs

The PAT scheme was started in 1983 in recognition of the deprivation caused to many people by lack of contact with animals. This can be a deprivation particularly felt by elderly people who, when taking up residence in care homes, are often not allowed to take their pets with them and sometimes even have to have their pets destroyed in order to comply with these rules. The scheme is supported by the 'Pro Dogs' charity and more than 9,500 dogs have now been registered as PAT dogs. They regularly visit hospitals, hospices and homes for the elderly, as well as schools for children with learning disabilities throughout Britain. All dogs are tested for appropriate temperament and must be vaccinated, wormed and 'clean' before they can be registered as PAT dogs. The scheme has become the largest community service in Europe where dogs bring benefit to people on a regular basis. As the late Lesley Scott-Ordish, founder of the PAT dog scheme, commented: 'People do not easily forgive themselves if forced into the situation of agreeing to have their much loved elderly dog destroyed through outside pressures and the guilt feeling may make it much harder [for them] to adapt to new and strange surroundings.'

PAT dogs offer predominantly elderly people respite from institutional life. They bring a touch of normality to often sterile environments and provide a link to the outside world. They also provide a distraction from the routine of hospital or nursing home life and visits are something to which residents can look forward. My own research focusing on the effects of a visiting PAT dog, Biba, found that in addition to the above, residents who had been pet owners before coming to a nursing home derived a similar degree of social support from the dog as they would from a visiting friend.

My findings were based on residents' and staff responses to over one hundred questionnaires, interviews and observations of Biba's interaction with those living and working at The Royal Star and Garter Home for ex-service men and women in southwest London. In some cases, residents of the home found Biba's company had additional benefits to those drawn from human interaction. This was illustrated during an interview with one of the more severely disabled residents. Alan was in his late fifties and completely paralysed below the neck. He controlled his wheelchair with his chin. His speech was restricted but he could generally make himself understood. He had a particular fondness for Biba and very much looked forward to her weekly visits. I conducted a tape recorded interview with him discussing his relationship with Biba and what he derived from her social calls. I understood most of his responses with the exception of one. Even though Alan repeated it several times, I couldn't catch one of the words. To be honest, I didn't feel the response was especially significant and was happy with those I could understand. That evening I listened to the tape and for some reason the tape had picked up the missing response more readily than my ears. I had asked him how he felt when Biba was with him and I had originally only been able to discern the first two words of his three-word reply, 'Happy, relaxed . . .'. The tape provided the complete response, 'Happy, relaxed, NORMAL'. Biba's response to him was completely unaffected by his condition because, unlike us humans, it was of no consequence to her.

Biba, an elderly mongrel bitch, and her companion human Marion, are rather special to The Royal Star and Garter Home (RSGH). The Home, as it's title infers, has Royal patronage and Biba can boast The Queen Mother amongst her many admirers. Marion and Biba have called on residents of the home for over ten years. They visit every Thursday and this gentle dog is available to residents to stroke, cuddle and feed with biscuits (at Marion's discretion). Biba was officially adopted as the RSGH's mascot and is often seen at fund-raising events in a coat bearing the Home's crest. Often on such occasions she is accompanied by residents.

Biba is always ready to help even when she is off duty. This letter was recently sent to me and describes events in a local doctors' waiting room:

'This is a pat on the back for a PAT dog! I wanted to tell you how much I appreciated Biba's assistance a few weeks ago . . . As a medical receptionist I have to sit with the wife of one of our patients while he has treatment, as she is in the early stages of senile dementia (let's call them John and Mary). John is deaf and therefore their daughter accompanies him into the treatment room, leaving Mary by herself with me. After a while Mary becomes restless, wondering and worrying where her husband and daughter have gone and wanting to go and look for them.

'Fortunately for me, on one such occasion Marion and Biba were there. I was very busy answering the phone, making appointments and receiving patients. Mary was encouraged to speak to Biba, who responded to the situation immediately, so much so, that although Marion left the room for her own treatment, Biba continued to remain on alert so that if Mary got up from her chair Biba would be there in front, taking Mary's interest and keeping her occupied. I even asked Biba to stay on when Marion was ready to leave as she had been so helpful. A proper PAT dog.'

My time researching at the RSGH was full of surprises. Given the Home's rather grand setting, overlooking the Thames Valley, and its seemingly institutional environment, I had not expected an animal assisted therapy (AAT) programme to be so readily accepted. On my first meeting with the Matron of the Home, I was overwhelmed by her enthusiasm for both AAT and my proposed research. She quickly explained that in one of her previous places of work she herself had arranged AAT on a rather large scale. She was working in a hospice at the time and one of the residents was in his last few weeks of life. He had spent much of his adult life in a circus working with elephants. He often referred to these happier times, so the Matron decided to arrange a rather special treat for him. She contacted a circus and they very willingly brought a baby elephant to visit him! Matron's enthusiasm for the PAT dog visits and AAT in general was shared by pretty much all of the senior medical staff.

The Royal Star and Garter Home was established as an independent charity in 1916 to care for the severely disabled sailors and soldiers returning from the battlefields of the First World War. Queen Mary and the British Red Cross were instrumental in setting up the Home, which was originally housed in the old Star and Garter Hotel on the present site. The old hotel was demolished in 1919 and a new purpose-built home erected. Designed by Sir Edwin Cooper, the new Home was opened by King George V and Queen Mary in July 1924 and dedicated as the Women of the Empire's Memorial of the Great War, the British Women's Hospital Committee having raised the funds to build it. During the construction period, a temporary residence on the south coast of England at Sandgate was used to accommodate the residents. This home was also extensively rebuilt and the two homes ran together until 1940, when it was thought prudent to close the Sandgate home due to the threat of invasion.

Today the Home cares for up to 185 disabled ex-service men and women (currently 160 men and 20 women), of all ranks and ages and from all three services, including reserves. Although originally for those disabled on active service, any ex-service man or woman is eligible for a place today, regardless of how their disability occurred. Residents today include those suffering from a range of disabilities and conditions such as strokes, multiple sclerosis, motor neurone disease and arthritis, as well as those disabled through serious accidents.

The eight wards or 'suites' provide almost all residents with single room accommodation, apart from a small number of twin-bedded rooms and open bays for acute cases, and include a sixteen-bed residential unit. Many of the residents are elderly and in need of intensive nursing. The Home is equipped with special beds, pressure relieving mattresses, hoists, special baths and other lifting apparatus. Respite care and periods of rehabilitation are available, as well as a permanent home for those who need it. The rehabilitation unit offers physiotherapy, hydrotherapy, occupational therapy and speech and language therapy. These facilities are available to every resident, short-term and long-term, and those who cannot leave their beds are treated on the ward.

The welfare department offers a wide range of leisure pursuits, both inside and outside the Home. Frequent concerts and entertainments are held in the large Queen's Room, which is fully equipped with stage, lights and sound system. The room is also used for indoor bowls, wheelchair dancing and social evenings. Regular outings and holidays feature throughout the year. An army of volunteers ensure the smooth running of the music, chess, and bridge clubs, the residents' bar, and the library. Volunteers also act as escorts on outings and help out on the wards and in the dining room. In addition, the Home also has its own pharmacy, dental surgery, chiropodist, counsellor, medical social worker, psychologist and hairdresser.

I shadowed Biba's Thursday morning rounds on the wards and was struck by the warm welcome she received from both residents and staff. Her placid nature and impeccable manners made her very hard to resist and many members of staff took a few minutes out of their very busy schedules for a stroke and cuddle. Biba's advanced years were reflected in her clouding eyes and greying coat but she seemed to take her duties in her stride. It was also very apparent that Marion, the 'companion human', was a welcome distraction too.

On one of the wards, while accompanying Biba and Marion, I was approached by an elderly lady who had very recently arrived at the Home for two week's respite care. She explained that she was a little anxious being a 'new girl' and was curious about my presence. I began to explain but before I could finish, she interrupted my flow, in a rather strained tone, with 'There's a dog here?' I was a little concerned because I was fully expecting a negative reaction. In my most reassuring of tones, I answered in the affirmative, but before I could get more than three words out she interrupted again with, 'Oh I'm so pleased. Animals, they make life so worthwhile, don't you think? I'm going to be happy here I can tell, for a while I was a little worried.' Marion introduced herself and Biba and informed the new resident of the days and times of her visits. It was clear that Biba had been instrumental in relieving this woman's initial anxiety and subsequently provided her with ongoing support throughout her stay at the RSGH.

Cathy, another elderly female resident, who had been at the Home for some time, had established a particularly close relationship with Biba and readily consented to a tape recorded interview. Cathy had a soft and lyrical Scottish accent and when we met she described her childhood home on the shore of a loch. Dogs had always been part of her life there. She spoke of a collie dog who had been her close companion as a child and of a special 'gift' the dog possessed. Her father worked on a loch steamer and this took him away from home for days at a time. The collie was also very close to her father and was able to predict his return from a steamer trip. The dog would suddenly leave any activity he was involved in and take himself down to the landing dock on the banks of the loch, where he would sit and wait for the steamer's arrival. Without fail, the small ship returned within two hours of the dog taking up his position.

At the time I suppose I put this story down to a rose-tinted reminiscence of a childhood long ago, but 18 months later I had cause to reflect upon it again. I was sitting, rather nervously, waiting to present a paper to a conference at Cambridge University. The conference was being addressed by one Rupert Sheldrake. His presentation was entitled 'Pets That Know When Their Owners Are Coming Home'. His findings were absolutely fascinating: a survey in the north of England revealed that 46 per cent of dog owners and 14 per cent of cat owners had noticed that their animals seem to know when a member of their house-hold was about to return, half an hour or more in advance. He had also conducted numerous experi-ments with his research partner Pamela Smart's dog, 'Jaytee'. These had been captured on video and clearly indicated that the dog had an ability to predict her owner's return. Jaytee repeatedly went to sit by a win-dow in Pat's flat at the point when she was about to leave different locations for home. The findings could not be explained in terms of routine and type of trans-port used, whether it be bus, car or taxi. Rupert Sheldrake concluded that, 'This anticipatory behaviour does not seem explicable in terms of normal sensory information'. Another one for Scully and Mulder, I guess.

There was one lone member of the senior medical

staff at RSGH who was less enthusiastic about AAT. This person felt that Biba's efforts were on a par with the 'additional services' offered by the Home, similar to the musical entertainment which took place on an occasional basis and the visiting library service. In addition, this person decreed that the keeping of even caged pets by residents in their rooms, was prohibited for hygiene reasons and because of the possibility of cross infection. This view was not shared by the other senior medical staff. One of my first experiences, as a researcher at the RSGH, very quickly brought home to me that one particular resident also had very strong views on the subject. Arriving for my first meeting with Matron, I was informed that she was on dining-room duty. I made my way to the dining area and on entering the crowded hall, I saw her on the other side of the room. As I made my way towards her, I realised that she was trying to placate a very angry man who was standing and waving his walking stick in the air and shouting. The other residents on the surrounding tables appeared a little concerned by his actions and it became increasingly apparent that she was dealing with a very tricky situation. I continued to walk towards her with a view to offering some assistance. It was at this point she noticed me and made it very clear I should leave the dining room and wait for her in her office.

I made my way to her office where she met up with me some 20 minutes or so later. She was very apologetic and indicated that the elderly gentleman's actions were not unrelated to the focus of my research. Matron explained further. Several of the residents ignored the rule about not keeping birds in their rooms. The vast majority of staff were happy to collude with the residents and said nothing of the birds' presence. Unfortunately, the birds in the room of the man waving his walking stick in the air, had been discovered by the member of the senior medical staff mentioned earlier, and had been removed against his wishes. This had upset him terribly. On seeing Matron, who he believed was in some way responsible for the loss of the birds, he became extremely angry, hence his behaviour in the dining room. It seemed so terribly unfair. However, my spies tell me that the 'no birds rule' has now been relaxed.

It was very clear from my time at the Royal Star and Garter Home that Biba and Marion made a significant contribution to the well-being of many of the Home's residents. During their regular weekly rounds they visit all of the wards, which enables residents who are bedridden to have a few moments with them. However, this was not always the case. When Marion first began visiting the home with Biba and Max, her other dog who is sadly no longer with us, their visits were restricted to the occupational therapy department. This situation changed when, about 18 months after the visits had commenced, Marion was approached by the Deputy Matron who asked if she and the dogs would follow her to a ward because they wanted to try an experiment. A gentleman in his late sixties, Robert, had been admitted who had great difficulty using his hands. The Deputy Matron requested that Marion bring one of the dogs to the new resident's bedside. Marion took Max, who was slightly larger than Biba. Immediately Robert began to slowly stroke Max. His response to Max was witnessed by several members of the care staff, including the Deputy Matron. After his meeting with Max, Robert seemed encouraged to try to use his hands as much as possible. Marion was soon informed that she was now free to visit the wards.

Marion firmly believes that one of the major benefits of her rounds with Biba is that their presence can often encourage communication. She has recounted that several residents who had been less than forthcoming with staff, or who had even refused to communicate at all, had been quite willing to talk to her and the dogs. As a consequence, she was able to relay important information gleamed from these conversations to staff.

Marion has a warm and rather lively relationship with many of the staff. Communications between them consists, primarily, of conversations laden with acid comments and a mutual exchange of good humoured insults – the language of firmly established friendship. When Marion visited with both Max and Biba, a certain welcoming salutation from the nursing staff always made her feel at home, 'Here comes the dog with two bitches!'

Marion visits a number of other nursing homes and hospitals in southwest London with Biba. She recently acquired another, younger dog, Sonny, to help with these duties since Biba has had a few problems with her health. Wherever she goes, Marion's manner and her impeccably behaved dogs provide a distraction and warmth so necessary to soften institutional and clinical environments. I think most residents and staff would agree that the Royal Star and Garter Home wouldn't be quite the same without them.

There are many colourful characters among the ranks of the PAT dog volunteers and on a recent trip to the northwest of England I had the pleasure to meet Val, who visits hospitals in the Manchester area with up to seven dogs at a time. As a young women Val won a scholarship to study to be a vet at the Liverpool Veterinary School, but unfortunately, after only 18 months, she had to give up her studies due to developing an allergy to certain types of fur. Soon after this disappointment she trained as a nurse and entered a career which was to last 40 years. She worked in both casualty and geriatrics before spending 23 years in midwifery. She married Vic, a serviceman, who had lost both legs in an explosion during the Second World War and had severely impaired vision and hearing and extensive paralysis. He spent five years recuperating in hospital and they met when he was having a piece of shrapnel removed. It was watching her husband's reaction to animals which drew her to become involved with animal assisted therapy. With a good deal of practical and emotional support from Val, Vic managed to ride her horse – Flare. Val described to me how her husband 'felt like a man' when he was on the horse since instead of looking up to people from his wheelchair he had a chance to look down on them. She also very quickly became aware that non-human animals did not judge people by their disabilities. Sadly, Vic died quite recently, but Val's enthusiasm for visiting hospital wards with her dogs is undiminished.

Val has two German shepherds, Pi and Perdy, plus Banner, a Polish lowland sheepdog and Polly, a working sheepdog. She takes all of the dogs on her visits to the hospital and they are often accompanied by her friend, Shirley, and her three chihuahuas. It must be quite a sight – Val, Shirley and seven dogs of all shapes and sizes doing their bedside rounds. I asked Val what effect she feels they have on the ward environment when they visit. Her response conjured up a picture in my mind of controlled chaos, 'Well I suppose I take the place by storm. I just walk in and say hi, here we are again.' Val feels that she makes a very positive impact and certainly brings many patients out of their shells.

The patients' reactions to this therapeutic dog team vary from stroking to talking about the dogs and, for some, this canine cacophony quite simply stimulates a sometimes long-lost interest in their living environment. In most cases Val believes that the patients will eventually react because their visits are so difficult to ignore. Long-stay patients invariably strike up a rapport with one of the dogs or with Val and Shirley, or the whole lot of them.

Occasionally people react negatively and remark that dogs are dirty and have no place being on a hospital ward. Val attempts to allay their fears by explaining that if the dogs are well looked after, as her dogs are, there is very little chance of any problems with hygiene. If they persist in their objections, she just agrees to differ and moves on. Generally, however, the hospital authorities and doctors are very positive about her time on the wards. Val described to me a potentially awkward moment when her dog team's visit coincided with a consultant's rounds. She immediately offered to move on to another ward, to which he replied, 'No, no, no. Don't let me disturb you.'

Val highlighted an aspect of PAT dog ward work which is often overlooked: hospitals are less than inviting environments for most people, so what must it be like for dogs? For many of us, one of the strongest sensory associations with hospitals is the all pervasive smell of disinfectant. I would guess that this alone could unsettle some dogs. This, in addition to the unusual sounds, the differing physical environment and activities, is a lot for dogs to contend with. However, Val feels that her dogs detect that she is very comfortable on the wards and this puts them at their ease. After 40 years as a nurse, a hospital ward is almost a second home to Val. Her experience also offers a valuable insight into the nursing staff's response to the PAT dog visits. She believes that the dogs are a welcome distraction from the stresses and strains of nursing, and provide a break in routine. On particularly stressful days, they may also provide an opportunity for a much needed cuddle.

Val has witnessed many occasions when her dogs have provided comfort for patients who have been unable to derive such support from elsewhere. She needed little prompting to recount two encounters which illustrate the unique role PAT dogs can play. When she first began visiting her local hospital, she came upon a man in his late forties called Jerry, on one of the longer-stay wards, who permanently lay in bed in the foetal position. Jerry was a good looking man who had been a senior pilot up until he became the victim of a brutal assault. He had been hit on the head and mugged in Manchester, sustaining severe brain damage as a result of the attack. He had recently been transferred from one of the main Manchester hospitals for long-term support and hospitalisation. He had no movement, speech or acknowledgement. After seeing him on a couple of visits, Val asked the Ward Sister if she thought he might like dogs. The Sister was unsure but agreed to ask his wife, Sally, who was coming in later that day to see him.

On her next visit the Sister told her that she had spoken to his wife and she had informed her that they used to have a spaniel and Jerry was very much a dog lover. On hearing this Val had an idea. She proposed to try and open Jerry's arms, which had remained rigid in the foetal position, and place Polly the sheepdog in them. After consulting with the Ward Sister, she set about enacting her plan. Val lowered the cot side and very carefully prised open his arms and said,

'Jerry, Polly the dog has come to see you', and placed her on the bed with him. Polly remained still, snuggled close to him and then, almost from nowhere, a smile spread over Jerry's face and he tried to stroke her forehead. As he began to smile with the dog in his arms, Sally arrived at his bedside. On seeing his reaction to the dog she broke down and sobbed. Through her tears she said to Val, 'That's the first anything. We've never had anything out of him until now.' Val couldn't contain her own emotion and soon she too was crying. As she recounted these events the tears returned.

Following the momentous events of their first meeting, Val and Polly made regular visits to Jerry every Thursday afternoon. When possible, Sally would join them and Val would leave Polly in a smiling Jerry's arms while she visited other patients. The visits continued for almost nine months when, sadly, Jerry died. Sally wrote to Val following his death, expressing her thanks because when Polly was with him, it seemed he had at least one thing he could understand.

The second encounter Val related described the effect of a one-off meeting on the wards. Early one afternoon Val and her seven charges were doing their rounds when she came upon a new patient who made her stop in her tracks. This patient, Joyce, a woman in her early forties, was the thinnest person Val had seen in 40 years of nursing. She had no idea what Joyce's diagnosis was but when the dogs passed her bed she became very excited and called out, 'I've always wanted to have a dog'. Val replied in a very characteristic manner, 'Oh you can have two if you want', and placed the two smallest dogs, Polly and Penny, a chihuahua, in the bedclothes with one on each arm. Val gave them the command to stay and went off to finish the rounds with the remaining five hounds. On her return, about an hour later, she found Joyce and the two dogs fast asleep on the bed. So, she very gently slid the dogs out from under the bedclothes doing her best not to disturb Joyce and then made her way home.

That evening, just before she settled down to sleep, Joyce told one of the nurses how wonderful it was to have a dog for the first time because she had always wanted one of her own. That very same night Joyce died peacefully in her sleep. When Val next returned to the ward, she was informed of Joyce's death and how, due to her visit and Polly and Penny's company, her last hours had been some of her happiest.

Val's many years as a nurse, and her time with both dogs and horses, have lead her to develop her own theory of why close contact with animals can be so beneficial. She describes it as 'cuddling syndrome'. Her theory originates from her time on children's wards in the 1950s. She spent many night duties tending to extremely sick young children. A number of these children were struggling for breath and very close to death from diphtheria. She hated losing a child and in some cases felt that if she held them through the night and willed them to live, they wouldn't die. Some children did survive and Val began to realise the importance of being held or cuddled. However, she feels cultural conventions, the stiff upper lip attitude in particular, often prevents such positive displays of affection between humans. She believes the warmth of sensitive physical contact to be vital for our well-being and this is what the animals provide. As she puts it:

'. . . just something to put your arms around, to think oh thank God somebody cares. Just love me for a minute. To reach out and touch something, something that's warm . . . and even men like to cuddle.

Riding therapy at the Fortune Centre
(courtesy David Miller/Fortune Centre of Riding Therapy)

'Vaulting' — a remedial session in the indoor school at the Fortune Centre of Riding Therapy (courtesy David Miller/Fortune Centre of Riding Therapy)

A therapeutic donkey ride (courtesy of The Elisabeth Svenson Trust for Children and Donkeys)

All in a day's work at the Fortune Centre (courtesy David Miller/Fortune Centre of Riding Therapy)

Elliot on wash day

Shopping with Bertie

A helping hand

(Photos courtesy of Canine Partners for Independence)

Craig gets the ultimate reward (courtesy of Bernie Graham)

Dr David Nathanson at work (courtesy of Dolphin Human Therapy, Florida, USA)

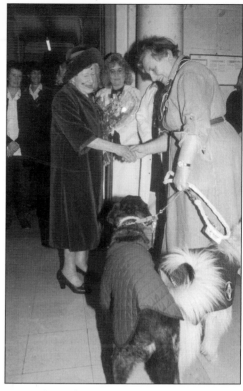

Simon Weston, a Vice President of The Royal Star and Garter Home, with Biba

Her Majesty the Queen Mother meets Marion, Biba and Sunny during a visit to the Home

Biba, the mascot of the Royal Star and Garter Home, with fellow PAT Dog Sunny (All photos courtesy of Mike Jones and The Royal Star and Garter Home)

Nora with Brillo — one of the PAT cats (courtesy of Lynn Spencer-Mills)

Val with her visiting PAT dogs (courtesy of *Manchester Evening News*)

Katie with a CHATA rabbit
(courtesy of CHATA)

Tiger cub Indy on a visit for CHATA
(courtesy of Jo Bomford)

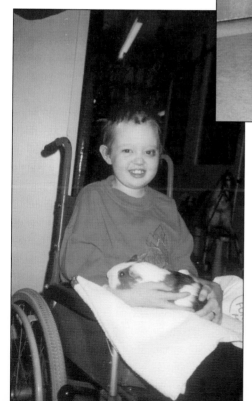

Jodie with Buttons the guinea pig
(courtesy of CHATA)

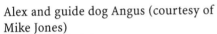

Trainee Dandy (courtesy of Hearing Dogs for Deaf People)

Morris Frank with Buddy, the guide dog trained for him by Mrs Dorothy Harrison Eustis in 1928 (courtesy of the Guide Dogs for the Blind Association)

Alex and guide dog Angus (courtesy of Mike Jones)

Family outing (courtesy of the Guide Dogs for the Blind Association)

These men in hospital, they love it when the dogs are on the bed and they put their arms around them and then they start to cry.'

For many people, she believes that the holding of an animal may turn back the clock to their childhood, when embracing, cuddling and physical contact in general, were not only more socially acceptable but an almost reflexive demonstration of comfort and support. However, for Val the comfort and support derived from some animals offers enhanced benefits because it is given so freely with little or nothing expected in return.

Besides their visiting PAT dog work, Val and some of her dogs are also involved in therapy sessions working with people who are almost disabled by their fear of dogs. Some of those who attend are unable to leave their homes due to their phobia/fear of meeting a dog. Val's German shepherd, Pi, has proved especially effective in this work. The logic is simple: if people can overcome their fears working with a breed that is often portrayed as particularly aggressive, this will stand them in good stead with dogs in general. Phobic individuals attend a 13-week desensitisation course. Initially, in small groups of three or four with the therapist, they enter a room where Pi is lying down. These sessions last about half an hour and the therapist asks the clients to gauge their level of fear on a scale of one to ten while in the dog's presence. There is no physical contact with Pi in the first few weeks but eventually, for those whose fear has diminished, there is the opportunity to stroke her or even take her for a walk with Val. At the end of the 13 weeks Val's other dogs are introduced. To date, the therapy has proved most successful.

Val gives regular talks on her work, which has been

reported both locally and internationally, and has addressed the Institute of Health at the University of Liverpool. In 1994 her work was positively evaluated in a study commissioned by the Japanese Ageing Research Centre.

Chance Encounters

I would like to describe a couple of episodes where the support and comfort derived from chance meetings with dogs has proved invaluable. I begin with an unexpected but never to be forgotten encounter which took place while I was researching that which you hold in your hands at this very moment.

My trip to Cheshire to meet Val meant a weekend stay in the Manchester area. This is familiar territory to me from times past and, stepping off the train in Stockport, I felt the presence of disturbed ghosts. Even though I was a regular visitor to these parts in my childhood and teens, I hadn't returned for nearly 20 years. I mention these details as a preface to describing some most unsettling moments I experienced during the weekend and how a brief encounter, with a very sensitive and supportive terrier, did much to ease my strained emotions at the finale of a painful journey into the past. I beg your indulgence for some background detail.

My mother, Frieda, arrived at Southampton docks in the summer of 1939, a nine-year-old Jewish refugee from Nazi Germany. She had travelled with her younger sister, Betty, from Frankfurt wearing labels which bore their names and refugee status. They were soon separated, with Betty being billeted with a family in London and my mother being sent to a spinster lady in Cheshire. Frieda travelled alone to the north of England on a train full of servicemen. She cried for almost the entire journey, despite the attempts of a kind group of sailors to comfort her. My mother has spoken very little about the following days, months and years, even though she remained in the region until her mid teens. Her father, my only grandparent who survived the horrors of the Holocaust, also eventually settled close by in Manchester.

During Frieda's wartime stay in Cheshire, she did, however, establish a friendship with Doreen Clarke which has lasted over 50 years. My mother had always told me of how good Doreen and her parents had been to her as a child, but she was reluctant to talk in any detail about this period in her life. Many of my earliest trips to this region were to stay with Doreen and her husband, Allen, while visiting my grandfather. For a brief period I courted their eldest daughter, Alison. When I arranged to meet with Val to discuss her work, it soon became apparent that she lived very close by to the Clarke's, so I decided to combine my research with a weekend in the company of these very dear family friends. This would be my first visit to their home for 17 years. I was very much looking forward to seeing them all again and, in particular, chewing on the past with Alison, who had recently returned to England with her two children, after many years in Canada. Alison's younger sister, Lindsay, had also recently had a baby and I had been invited to the christening, which was to take place on the weekend of my visit.

I've always found that travels into the past are as much about confronting that which has remained unaltered, as that which has changed. For me, in returning to these parts, the unaltered primarily comprised of feelings of unease about my mother's childhood, and the need to come to terms with reminders of the murder of most of my family from the previous generation in the Nazi extermination camps.

As the weekend progressed my emotions began to polarise between the delight of immersing myself in proven and valued friendship, and the despair of confronting memories of demons so despicable that it was dangerous to look upon them for too long. A chance conversation with Alison late on Saturday afternoon straddled these poles when she told me the full story of how her mother, Doreen, first met mine. We were sitting in her mother's lounge with Olivia, Alison's nine-year-old daughter, as she described the meeting that had taken place almost 60 years before. My mother was brought into Doreen's classroom by a teacher quite late in the summer term of 1939. She explained that Frieda was a Jewish refugee from Germany and could speak no English. The teacher then asked if anyone would like to help her settle in, to which Doreen immediately replied, 'I'll be her friend, Miss'. I don't know whether it was the impact of at last hearing details from this time, until now shrouded by my mother's denial, for fear of being consumed by an avalanche of unwanted memories, or the presence of a nine-year-old girl, to so vividly accent the account, but for a brief moment, and for the first time, I felt consumed by the pain of my mother's childhood years.

The next 24 hours prior to my return home became an internal rearguard action to fend off potentially overwhelming emotions. I couldn't just up and leave, as part of me so very much wanted to do, because that felt too much like defeat. So, as planned, I took the three o'clock train on Sunday back to London. I tried to distract myself by reading but my concentration had deserted me. I arrived at Euston station in the early evening and caught a connection to Willesden Junction. From here the final leg of my journey was a 15-minute ride to Kew Gardens. As I boarded the train, on auto pilot and almost oblivious to my surroundings, I felt that my emotions were finally about to get the better of me and I would be unable to hold back the tears any longer. I firmly believe that we should all cry whenever we need to, but for me this has always been the most private of activities. In a last attempt to stave off an unwanted public display, I closed my eyes, bowed my head and took a very deep breath. As I opened them, my eyes were met by those belonging to a most concerned looking black and tan Airedale terrier. The dog was sitting very still directly in front of me with his head slightly tilted to one side. He was with a young women in her mid twenties who was deeply immersed in her book. I dispensed with my usual enquiries regarding the dog's name and nature and if it was all right to stroke him because, not only did I not want to disturb her but the lump in my throat had become so large by now that speech would not have come easily. I slowly lent forward reaching towards the dog and gently stroked the side of his head. He very tenderly pressed his head against my hand, while continuing to observe me with eyes which conveyed a seemingly knowing caring and concern. His fur was warm and surprisingly soft. His unquestionably calming presence was slowly pulling me back from my emotional precipice. I continued to stroke him and he very delicately licked my hand. I was still full up inside but his reassuring support was enabling me to contain my feelings until I was ready to let them flow. As if to seal my survival, he stood up, pressed his head to my thigh, and then lay down upon and across my feet. And that is where he remained until he and his companion human left the train one stop before my own. As they got up to leave I gave him a final cuddle and I swore he smiled. Just before the automatic doors opened to allow them to leave the train, he gave me a final glance over his shoulder and then he was gone. I eventually returned home and in my own room, in my own space, and the arms of Patsy, my wife, I was able to let go.

Chance meetings with dogs can also have the most far-reaching effect, as the following account so clearly illustrates. Cathy, the wife of a very old school friend, is the Matron of a small nursing home for elderly people in west London. She recounted the following episode concerning one of her residents and I must commend her for adopting such a flexible approach to working with individuals, rather than the more common broad-brush approach to care. Connie, a woman in her early seventies came to live at the home under somewhat difficult circumstances. She had a 40-year history of severe depression and had been in hospital on many occasions. She usually responded to treatment but when Cathy first met her in hospital she was particularly low and withdrawn and was becoming increasingly paranoid. Connie was convinced that she was wanted by the police and that she owed vast sums of money, neither of which were true. On this occasion, the efforts of the medical staff appeared to be in vain. She was not responding to treatment and the depression was becoming entrenched. However, she was still capable of washing, dressing and feeding herself, so Cathy had no hesitation in offering her a room at the nursing home.

None of the medical professionals involved in Connie's care were aware of any life event that may have triggered the almost continuous depression that had plagued most of her life. The information that was available suggested that she had always been a loner and had spent her working life as a waitress. Other than these scant details, very little was known about her past. However, Cathy and her staff were prepared to work with her and offer any support they could. Connie spent the first month or so in her new home in a very withdrawn state. She didn't want to talk to anyone and only spoke if she really had to. She would spend her days sitting with her head bowed and her eyes closed.

This situation continued until chance played its hand. Jeff, the chef at the nursing home, had a small Pekinese dog called Hanjy. On one very harassed day, he had no alternative but to bring Hanjy to the home with him. The dog's presence was not a regular feature of the home's environment and on entering one of the main sitting rooms alone, he excitedly waddled around sniffing the residents and looking for strokes. When he got to Connie he suddenly stopped. For the first time since moving into the home, she lifted her head, smiled, put her hand down and stroked the dog. A somewhat shocked Cathy and a rather moved staff team held an impromptu meeting in the kitchen. Everyone agreed that the continued presence of the dog might break the ice with Connie and as a result they hatched a rather unusual care plan. They asked Jeff to bring Hanjy in every day for a week and this he did. Hanjy seemed to gravitate towards Connie. She seemed pleased with his company and, as a result, was slowly becoming less withdrawn. The following week they asked Jeff to go into the lounge with Hanjy and play with him. Jeff was very pleased to do this because he was very fond of the residents. The two of them romped about the floor and Hanjy performed all his tricks. He seemed to put on a special effort to perform when Connie was present.

Hanjy continued to visit over the next week or so. Jeff would pop into the lounge to see if the dog was behaving and he soon began to develop a rapport with Connie. Their conversations initially revolved around Hanjy but Cathy had encouraged Jeff to sensitively introduce other topics. Connie responded with some interest and also began talking about herself a little. As part of a very well intentioned set up, Jeff discussed the kitchen and then, at a pre-planned time, he told her that he had to get back to the kitchen because of a staffing problem. He explained that he was so short staffed, that he would have to lay the trays and tables himself, in addition to his own work. Well, the plan worked because without any prompting, Connie said, 'I can set trays, I can lay tables', and offered to help. This was just the beginning. She began to help on a regular basis and started clearing up after meals and chatting with staff while washing up with them. Staff also became aware that she was watching them help residents at meal times who couldn't feed themselves. Without a word to anyone, she too began helping others with their food.

Even though her day-to-day life at the home had improved, Connie remained reluctant to venture outside for fear of being arrested by the police. However, with a little help from a certain Pekinese, she is now beginning to overcome her reticence. Staff, by encouraging her to see what Hanjy is up to, have persuaded her to go into the garden. At the appropriate time, in the not too distant future, they are hoping that she will feel secure enough to take Hanjy for a walk to the end of the road.

Cathy is convinced that Hanjy has been the key to Connie's recovery and the subsequent improvements in her quality of life. Her social worker has been quite stunned by the transformation. She had never even seen her smile prior to Hanjy coming on to the scene. Cathy and the social worker acknowledge that she still has obstacles to overcome but for a women who, in the past, was so withdrawn that she only occasionally opened her eyes, rarely raised her head off her chest and ended conversations by just walking away, she has come an awfully long way.

CHATA – Children in Hospital and Animal Therapy Association

CHATA is a registered children's charity founded six years ago by Sandra Stone. It is staffed entirely by volunteers and is run by Sandra and her husband Ronnie from their home. The main aim of the association is to improve the quality of life for seriously sick and terminally ill children by offering animals as therapy. Animals used include dogs, rabbits and guinea pigs. CHATA currently attends eight London hospitals, visiting children in almost all areas including intensive care. All volunteers work with their own animals and are medical professionals or trained to work with children. The animals used are vigorously checked by a veterinary surgeon for transmittable diseases and assessed for docile temperament. CHATA maintains a close relationship with the hospitals' infection control departments.

Apart from the everyday work with domestic animals, CHATA have links with wildlife parks and farms. More exotic animals, including tiger and lion cubs, have been introduced into more controlled areas of the hospitals. This adds excitement but is carefully monitored. Unlike the sessions with domestic animals there is no physical interaction. The children watch, transfixed, and ask questions.

CHATA take animals into hospitals because they are fun and provide a touch of 'normality' to the strangeness of the hospital environment and a link to the outside world. Animals are non-judgmental and can often help depressed and distressed children to communicate. Boris Levinson, a psychotherapist, was the first to document a pet animal's potential to ease children's communication in hospital or treatment settings in the early 1950s. His first observations were drawn from a chance event when a very withdrawn child he was treating arrived early for an appointment. Levinson's dog, Jingles, happened to be in the office with him and it quickly became apparent that the child, who interacted in a very positive way with the dog, was less distant and more inclined to communicate. Jingles seemingly acted as an intermediary by breaking the ice and enabling the child to feel more at ease in the treatment environment. The CHATA animals work in a very similar way. They appear to make it easier for a child to show and express feelings of affection. Additionally, they offer children an opportunity to take responsibility and be carers rather than the cared for and, very importantly, provide a distraction from illness, hospital routines and the clinical environment. A scientific study is presently being conducted to evaluate the benefits of CHATA's work. This research has been commissioned by Dr Ian Pollock, a Consultant Paediatrician, in the light of an abundance of persuasive anecdotal evidence.

CHATA also organise days out for children able to

leave the hospital to visit wildlife parks, arrange for disabled children to attend therapeutic riding sessions and, where possible, enable them to swim with wild dolphins.

In April 1998, I attended a conference on animal assisted therapy jointly convened by CHATA and the Society for Companion Animal Studies (SCAS). The conference was held at Guy's Hospital, London, and focused on raising professional and public awareness of CHATA's work. The initial findings from hospital-based research on the benefits of CHATA's visits were also presented. I knew many of the people attending and was very much looking forward to hearing the speakers, many of whom were experts in their field. One of the names on the list of speakers, Sue Markham, was unfamiliar. This remarkably brave women spoke only briefly of her recent experiences but made such a deep impact on me that I immediately knew I wanted to include her in this book. With the help of Sandra Stone of CHATA, an interview was arranged. This is Sue's story.

Sue's son Stephen was born with heart disease. He was operated on at the age of ten months and it appeared that the procedure had been successful. However, during a family trip to Australia when Stephen was eight, it became apparent that there was a problem with his breathing. On their return to England, Sue took him to see a consultant at Harefield Hospital and was informed that at some point in the future he would definitely need a heart and lung transplant. They were assured that there was no immediate cause for concern and were encouraged not to worry because the operation wouldn't be necessary for some time.

In March 1994, when Stephen was thirteen, he was

called into hospital for the transplant. Although the operation went well, he was never really himself again. He began to experience mood swings and was unable to return to school on a regular basis. He also lost a huge amount of weight and was being treated by a metabolic surgeon at Great Ormond Street hospital. In February 1996, just prior to an appointment, Sue became alarmed at his condition and contacted the surgeon and expressed her concerns. She explained that she was sure that once the doctor saw Stephen, he would admit him. The doctor felt this was unlikely, but when he did meet with Stephen he realised that his condition had indeed deteriorated, and he arranged for an immediate admission. Unfortunately, after two weeks on the ward, Stephen developed septicaemia and was rushed to the intensive care unit. His chances of survival were considered no more than 50/50. However, after six weeks in the unit, he pulled through and was returned to a general ward.

After a couple of days back on the ward, Sue took Stephen for a ride in a wheelchair and in the reception area of the hospital they saw two ladies, one clutching a rabbit and the other accompanied by a large Labrador. Stephen immediately wanted to go over and say hello and it soon transpired that the two ladies were Sandra and Hilary Summerfield from CHATA. They introduced themselves and told Stephen they would be very happy to come and see him on the ward if he would like it. Sue was very much taken with the idea and, from then on, following this chance meeting, they came to visit Stephen every week, during which Stephen became very attached to the animals, especially Dylan, the Labrador.

Stephen, who was now fifteen years old, was still experiencing extreme mood swings and on the second occasion he met Sandra and Hilary he had become very withdrawn and wouldn't talk to anybody. He refused to speak to Sue or his father or the nurses. All he seemed to do was scream. Sandra, a former nurse, and Hilary, a retired GP, were not deterred however. As they approached Stephen he nodded an acknowledgement. He then sat and stroked the animals while he talked to them and by the time they had left he was considerably calmer. Sandra and Hilary continued to visit him over the following weeks and it became apparent that he was calmer and in better humour when the animals were around. CHATA organised a visit to an animal park for the children in Great Ormond Street Hospital. Stephen was invited and was allowed to stroke some young tiger cubs. This was a wonderful experience for him, as his mother, Sue, acknowledged: 'Whether it had any benefit on his health I don't know but he was certainly tickled pink, he was absolutely delighted to be able to do it.'

It was now July 1996 and Stephen was ready to leave hospital. He was begging for a dog but Sue couldn't persuade her husband to agree to it. Sue hadn't given up on changing her husband's mind but she was sure getting a Labrador was too much to expect. She approached Sandra for some advice on a more suitable dog. She suggested a smaller dog, a lhasa apso and then brought her own, Fifi, to the hospital for Stephen to meet. He immediately fell in love with her. So it was decided that they would try to get one and even Stephen's father warmed to the idea. When Stephen left the hospital, Sue contacted the Kennel Club and within a fairly short space of time they managed to acquire a six-month-old bitch. Stephen named her Bandit and very quickly they became inseparable. When Stephen went out in the wheelchair, she either sat on his lap or rushed beside him. If he had to stay in bed, which he did for a considerable amount of time at this stage, she just lay by his side all day and didn't move. Stephen was calmer when Bandit was with him, but he was still given to extreme mood swings. He lashed out at his family on occasions, usually verbally but sometimes physically, through absolute frustration. However, he never did this to the dog.

Stephen's condition deteriorated and he began to suffer an enormous amount of pain. He couldn't bear to be touched and Bandit seemed to sense this. She would rush enthusiastically at everyone but him. She would just lay gently at his side. Before too long Stephen became desperately ill and, sadly, he died on the morning of 4 July 1997. Bandit had been with him almost to the end. For the Markhams, as for any family, the pain of losing their son has been extremely difficult to bear. Bandit clearly shared their sense of loss. She was devastated and for many weeks pined for her companion.

Within a few months of Stephen's death, Sue became an active CHATA volunteer. She had been convinced of the benefits of CHATA's visits by Stephen's and other children's responses to the animals. She told me of a little two-year-old boy in the room next to Stephen's. He had had surgery on his throat and had never spoken. He seemed very bright and could make himself understood very well with his hands. Sue described to me how ecstatic he became when the animals came to visit. In her experience, all the children she saw were thrilled to bits when the animals were on the wards. Bandit is now very much herself again and, following the completion of a range of suitability tests, she has made a couple of visits to the children's wards with Sue as a CHATA dog and by all accounts has thoroughly enjoyed the experience.

Sue spoke at the CHATA/SCAS conference within a year of her son's death. To be honest, I don't think in the circumstances I would have had it in me to address such a large number of people. Her courage bears testament to both her love for Stephen, and her belief in the comfort and support animals bring to children who have to face such anguish so early in life.

Current Developments

The two following organisations appear to be at the cutting edge of assistance dog work and training. I'll add little else to this introductory paragraph because their activities speak very much for themselves.

Support Dogs and Seizure Alert Dogs

Support Dogs was set up in 1992 to train dogs to assist their owners with their specific disability. Support Dogs differs from Dogs for the Disabled in the respect that they utilise the already existing bond between a dog and his/her owner. The dog and owner work as a team from the beginning of the training and therefore they both learn and gain confidence together. The training depends on the dogs knowing their homes and their owners before beginning to learn their special skills.

Support Dogs trains any type of dog, provided their temperament is suitable for the pressures of public work. Prior to training, each dog has a full behavioural and veterinary evaluation to ensure that they are mentally and physically capable of undertaking the work.

Initial training includes ensuring that the dog is comfortable at home, knows basic commands and the owner has control of the dog. Following this the dog is trained/socialised in a variety of situations to enable him/her to work in public areas. Support Dogs are allowed entry into restaurants, food shops and other food premises because their special training means they are not a risk to hygiene in such places.

Advanced training includes intensive residential work at the main centre in Sheffield, as well as continuation of the training in the dog and owner's home. After a minimum of 170 hours of instruction and a final assessment the dogs qualify as Support Dogs. Each one is trained for his/her owner's specific needs. The tasks they undertake are very similar to those of Dogs for the Disabled.

Support Dogs have recently extended their activities to work with dogs who are able to predict and alert their owners to an imminent epileptic seizure. These dogs are trained alongside the person with epilepsy as they have to respond to that individual's specific type of seizure activity. The group's work in this area began when they were approached by a young woman, Tony Brown-Griffin, who had noticed that her five- year-old Border collie, Rupert, seemed to have the ability to predict and alert her to an imminent epileptic seizure. She described her experiences with Rupert in a TV interview broadcast throughout Britain in 1998. Tony explained that she has two types of seizure up to ten times a week. She described the effects of the first as having a feeling of mental absence, as though her mind had switched off. The second was the more recognised major epileptic seizure.

Her husband was close to giving up work to care for Tony when they first became aware of Rupert's ability.

They contacted Support Dogs to see if they could develop and train this ability. They advised Tony to keep a video diary of Rupert's behaviour in relation to her seizures. It soon became apparent that Rupert could not only alert her to a seizure approximately 45 minutes in advance, but he could also differentiate between the two types of seizure. If Tony was about to have a major episode, Rupert would put his paws on her thighs and bark. To alert her to the second type, he would put his paws on her thighs or almost climb into her lap but without barking. Using the video diary, his specific alerting behaviour could be clearly identified and reinforced in training. To date Rupert has a 100 per cent record for correctly alerting Tony.

Rupert has had a major effect on the lives of Tony and her husband. The advance warning of a seizure can enable Tony to get to a place of safety. Before Rupert was part of the family she was very wary of going out. As Tony has said herself, 'Rupert has given me my life back.' If she is out and Rupert senses an imminent seizure, he will walk in front of her, block her path or turn away from crossing roads. She believes that he has saved her from being run over on at least three occasions.

The potential benefits for the thousands of people who experience epilepsy are enormous. In addition to finding a place of safety, advance warning of a seizure may enable a sufferer to take medication to reduce the severity of an episode or even prevent it.

To date, no one has established how dogs are able to predict human epileptic seizures but two possible explanations have been put forward: *1)* certain dogs may pick up on changes of mood or behaviour prior to a seizure; and *2)* a dog's extremely powerful sense of smell may be a significant factor – prior to a seizure

chemical changes take place and pheromones are produced in a person's sweat. A dog may be able detect
this change in scent or smell. A pilot study to investigate this intriguing phenomenon began in December
1997 at the University of Florida. The researchers
hope to ascertain clues from the study as to what it is
the dogs are responding to; how to help service dog
trainers to be more successful in selecting dogs for this
type of task; and how to help trainers and health care
providers decide which people could benefit most
from seizure alert dogs. At the time of writing, the
study was yet to be completed but one set of preliminary findings is particularly noteworthy. A number of
American seizure alert dog trainers were contacted
and questioned on their experiences. A consensus
appeared to emerge which suggests that seizure alerting is an innate ability not a trainable one; nor is it
breed specific, so any breed of dog or mix can be born
with the ability; dogs alert 50–70 per cent of the time
with a certain type of behaviour. People with alert dogs
reported that the quality of their lives had improved
since acquiring the dog and this view was supported by
family members and carers.

Support Dogs are currently researching seizure
alert dogs' abilities in conjunction with the David
Lewis Centre for Epilepsy. In contrast to the findings
from Florida, their experiences indicate that with specialist training any type of dog can be trained to alert.
However, they feel that as a result of their initial
research, they are now able to identify traits in some
dogs which may make them more responsive to
seizure activity. To date Support Dogs have trained
eight seizure alert dogs, with several more undergoing
training.

The research in this field is in its infancy and,

whether dogs can be trained or this sense is innate, it is clear that some dogs have exhibited an intriguing potential for detection and alerting which may offer support and hope to millions of people worldwide.

Canine Partners for Independence (CPI)

CPI train mainly golden retrievers and Labradors to help severely disabled people. CPI dogs respond to over 90 verbal commands and are able to:

- Select goods from a supermarket shelf, put them in a basket, stack them on the conveyor belt and pay the cashier.
- Call a lift, open and shut doors, and fetch items when required.
- Draw the curtains and take the washing out of a washing machine.
- Pick up dropped items and give them to their disabled owner.
- Draw a pension from the post office or withdraw cash from a bank.

CPI is very much a developing organisation and spearheading much of their work is one Nina Bondarenko, the Training Programme Director. Nina, an Australian, has been training animals, mostly dogs, for over 25 years. Before moving to Europe, she presented a talkback radio show on animal selection and behaviour, and a segment on a weekly Australian national TV programme dealing with all aspects of companion animals. She is a lively, effervescent character who certainly leaves a lasting impression.

Nina has been one of the pioneers of a differing

approach to the training of assistance dogs, derived
from her many years of observing canine behaviour. In
particular, she has drawn from her experiences working with Rottweiler puppies and the lessons learnt
attempting to train a dingo (Australian wild dog) from
a very young age. Her method differs from the more
established approaches in two respects: she begins
working with puppies at a much younger age, from as
early as five weeks old; and she attempts to train or
teach dogs in a way that is similar to how they would
naturally acquire a skill. As a result Nina believes dogs
learn more easily. More accepted human oriented
training methods, in her opinion, don't make a lot of
sense to a dog and make it harder for them to learn.
She feels her approach encourages dogs to problem
solve in addition to learning individual skills. Her
ideas appear to be catching on. Other assistance dog
groups are also now training puppies from an earlier
age. CPI dogs are trained to offer their human partners both functional and practical support but, as I
found out from Nina, the placement of a dog can have
unexpected results.

Tina, a young woman in her mid twenties suffering
from cerebral palsy, had applied for a dog but, on her
arrival at the CPI centre for two weeks' training, a
problem soon became evident. Due to her condition,
her speech was quite slurred and it was very difficult to
distinguish between the beginnings and ends of words.
She had undergone many years of speech therapy but
it was still far from easy to understand her. Nina was
concerned that the lack of clarity was likely to make it
very hard for a dog to understand Tina's commands.
This did, indeed, prove to be the case, but Nina felt
there was one dog, Destry, an 18-month-old golden
retriever, who might just be attentive and observant

enough to do the job. She wasn't too optimistic but felt it was certainly worth a try.

At first things didn't seem too hopeful. The dog was trying very hard to understand Tina but was getting confused and becoming upset. Tina also became very frustrated and felt terribly bad for the dog. However, within a couple of days, Destry began to understand her slurred speech and they were soon working quite well together. The outlook now seemed a lot brighter, but Destry had to understand quite a range of commands if they were going to pass the final test. If they failed, Tina could not have the dog. Sometimes, with help from above, finding the right motivation can do the trick. By the end of the training period a near miraculous transformation had taken place. Tina's speech had improved almost beyond belief. She had been so determined to speak more clearly so Destry could understand her, that two weeks of training with a dog had achieved what years of traditional speech therapy had failed to do. Not only had her speech improved, but when she was filmed taking her final test, she had the clearest voice, the best phrasing and the most controlled working relationship with a dog of all the trainees.

Another young golden retriever, Danny, was placed with Tom, a man in his forties who had multiple sclerosis. Tom's wife, Karen, was also disabled following a massive stroke some 15 years earlier. Danny had been placed with Tom to pick up items and accompany him on shopping trips. He was also there to help Tom support Karen, who had extensive paralysis from the stroke. She could walk a little with the aid of a stick, but spent most of her time in a wheelchair. Her speech was restricted to just a few words – 'yes' 'no' and 'coffee'. She had become very depressed over the years, and as

Nina recalls, Tom was the only person who could sometimes understand her efforts at speaking. Her more usual means of communication with Tom was to look at him and then hit him and then hit the thing she wanted, or to point at it. Her depth of despair and level of frustration had led her to attempt suicide on more than one occasion.

Even in these very difficult circumstances, Tom was very happy with Danny and told Nina what a wonderfully helpful dog he was. He also reported that Danny seemed to be trying to communicate with Karen and that she was trying desperately hard to respond but with little joy. However, as time went on, she interacted with Danny more and more and, to Tom's surprise, began playing with him. Then one morning about six weeks after Danny had been placed with the couple, Tom met with Nina and described the scene he had just witnessed prior to leaving home. Danny was running around the backyard with a big blob of grass on his head. Karen had found this terribly funny and for the first time in 15 years she'd been laughing out loud. This proved to be something of a catalyst for Karen. Following this event she slowly but surely began speaking again. She quickly let it be known that she wanted to train with Danny too. At first Nina gave her a clicker and a whistle for training purposes because her speech was still restricted. Karen then began working closely with Danny and giving verbal reports to Nina. Karen's response to Danny was completely unexpected. Nina explained to me that it is very unusual for people to regain speech more than a year after their stroke.

Nina firmly believes that CPI dogs placed with disabled people gradually tune into their new human partners and are sensitive to changes in their moods, emotions and behaviours. Similarly, she feels a bond also grows between the human and canine partners and the dogs develop 'context awareness'. This awareness is borne out of experience of their human's changing needs. I think the following account illustrates these points very well.

Nina placed a four-year-old golden retriever, Alfred, with Luke, a man with quite advanced multiple sclerosis. Luke's hands were very weak. On one particularly bad day, he couldn't take the weight of his purse and just kept dropping it. Alfred had repeatedly picked up the purse and put it back in Luke's hand but he dropped it again. This purse problem had been going on for some time and Luke was getting angry with himself. Alfred had become a little frustrated too. They had reached a point where Alfred was sitting with the purse in his mouth looking at Luke, and Luke was just sitting in his wheelchair staring at Alfred. Then from nowhere, Alfred leapt on to Luke's lap and pushed the purse in his mouth. Luke was a little stunned but absolutely delighted. This was a perfect solution. He could take the weight of the purse in his mouth and then just take money out when he needed it. From then on Alfred always put the purse in Luke's mouth. Nina admitted when recounting this tale that she would never have thought of training a dog to do such a thing. The dog had quite clearly worked it out himself.

Alfred's remarkable exploits didn't end there. Luke was in the bathroom one morning and dropped a jug that he needed for his catheterising process. He was alone apart from Alfred and his carer wasn't due for a couple of hours. At first he just sat there wondering what he was going to do. He didn't think to call Alfred because the jug was so big that he really didn't believe he could do much to help. Luke was also sitting with his back to the door and he couldn't really leave the bathroom until the catheterising process was finished. He was very close to resigning himself to being stuck and waiting for his carer. As he sat there he began to wonder where Alfred was and whether he may be able to help after all, when suddenly he heard him coming into the bathroom. Once inside, Alfred picked up the jug and stuffed it in the arm of Luke's wheelchair where it wouldn't fall out and then he just left, went back to the rug and lay down. Luke hadn't uttered a word to him but Alfred, without any fuss, once again resolved the situation. Luke, to say the least, was a little stunned.

It wasn't long before Alfred again came to Luke's rescue in the bathroom. When Luke needed to meet calls of nature or go for a wash, he always kept the door slightly ajar in case of any problems. However, on one visit, the door shut behind him and he couldn't get out. On this occasion he wasn't alone in the house but, due to his weakened condition, he was unable to shout very loudly and his family couldn't hear him over the TV. Alfred was soon on the scene but he was unable to open the door from the outside. So, Lassie style, he ran to the room where the family was gathered and barked at them until they realised there was a problem. Once he had their attention, he ran to the bathroom. They quickly followed and were able to let Luke out. The story didn't end there. From that day on, Alfred stubbornly refused to leave Luke alone in the bathroom.

The CPI trained dogs are quite adept at delivering the money to pay for purchases in a variety of environments and one dog, Alex, serves his partner particularly well in the evening by buying the drinks in the local pub. This skill is one of many which the canine and human partners develop together over two weeks of intensive training. The final test takes place in a local branch of Marks & Spencer. Nina, semi-hidden in the store, evaluates their joint working, focusing specifically on the human recipient's ability to look after the dog. It was on one such occasion that she overheard a rather stunned shopper turn to her friend and exclaim, 'Am I going mad or did I just see a dog buying a pair of socks!'

CHAPTER 6

HORSES THAT HEAL

Due to a painful childhood experience, this chapter is tinged with just a little personal irony. However, I am pleased to report that my adult encounters with these most noble of creatures have been rather more fulfilling. Nevertheless, my earliest recollections may offer a realistic and necessary perspective for considering equestrian therapy.

I spent the early years of my life in Hemel Hempstead. This semi-rural environment was one of a number of new towns built in England following the Second World War. Hemel, as it is more commonly known, is situated about 30 miles north west of London in Hertfordshire, and in my day, it had a population of about 50,000. Acres of fields, parks and woodland were to be found within 15 minutes' walking distance of my house and I remember several summers spent with friends making camps, playing football or cricket and seeing the horses. This last activity involved a less than 10-minute hike to an open field which was home to a varying number of horses and ponies. The area was securely fenced and our parents felt it a safe place for us to visit unaccompanied – once we had reached that milestone of maturity, seven years old. For most of my friends, seeing the horses came a poor fourth to camping, football and cricket but I thoroughly enjoyed it and so did one of my best friends, David. To a young child the horses looked enormous but I was adamant that, when I was just a little older and bigger, I would learn to ride. For now however, patting them, feeding them the odd carrot and watching them gallop, when the mood took them, was excitement enough.

David and I made regular trips to see the horses and

during the summer holidays we were very rarely alone on our visits. It was quite usual for six or seven children, aged from about five to 15, to be gathered enjoying the company and the carrot- and sugar-bribed attentions of the horses. Many of the older children had spent their earliest years in the bombed out or decaying areas of inner London. In retrospect, it seems that their gentleness with the horses reflected an appreciation of the long denied opportunity to experience the natural environment. In the main, the younger children were also very careful and sensitive. The horses were quite amenable to our advances and would only offer a friendly nudge if they were less than happy with any of our overtures.

We were certainly aware that these animals were exceedingly strong and powerful and that, if provoked, they could bite. So even though we used to sit ourselves on top of the wooden fence to get as close as possible to the horses, we never ventured into the field itself. No one ever attempted to mount or ride the horses, not until the day when 'the girl with no name' appeared. She must have been all of twelve years old. She was unusually stocky and was very much what many would call a tomboy. She had an intimidating stare and a less than friendly demeanour. David and I had never seen her before. She didn't seem to know any of the others so we tried to ignore her because quite honestly she frightened us. We were only about eight years old at the time so she was considerably bigger than us. One or two of the younger children took flight as soon as they saw her, which freed up a space on the fence next to me on my left. David was to the right of me, sitting balanced on the top of the fence. She quickly climbed up and filled the vacated place and I noticed she was carrying a stick of about 45cm

(18 inches) long. There were two largish horses very close to us, a chestnut and a grey. The chestnut was standing almost parallel with the fence, head down feasting on the grass. Our girl with no name turned to me and said, 'I'm going to ride 'im.' I can't remember my exact words, but I made it clear that I didn't think she should do it. She just told me to 'shut up' and struggled to climb from the fence on to the horse. He didn't appear keen on the idea and edged away a little. She made several unsuccessful attempts to mount him but eventually decided to give up. She was now back on the fence next to me and was quite audibly expressing her frustrations. She was cursing and shouting at her unwilling mount. The horse was becoming increasingly unsettled by her behaviour and the whole situation was becoming quite frightening. David and I were about to leave but it was too late. The girl with no name began hitting the horse with the stick. We shouted at her to stop but to no avail. In a desperate attempt to stop her I tried to grab the stick but she fought me off and pushed me away. As I reached to stop myself falling off the fence, I knocked into David who lost his balance and tumbled into the field. He stood up quickly and thankfully he was OK. However, David had landed within a metre (2 or 3 feet) of the hind quarters of a now very distressed horse. David couldn't have seen it coming but, as he took his first step to leave the field, the horse kicked out and caught him in the lower back. I don't think I'll ever forget David's torturous cry as the hoof hit. This time he stayed down, sobbing in unbearable pain. Those who had witnessed the kick were stunned into silence. The horse ambled away and I was able to get to David. He was in agony and sobbing uncontrollably. He could just about walk and with the help of some of the other

children we got him out of the field. I needed to get him home but I could not stop myself flying at the girl. I screamed 'this is all your fault' as I tried to grab her. She was too strong for me though and she threw me to the ground punching and kicking me. She was yelling that I had caused it to happen. I was now in tears from shock and her beating. Some of the other children pulled her off and made it very clear to her that she was to blame. I went back to David and helped him up. Almost hysterical, we very slowly we made our way back to his house.

David was very fortunate because no permanent damage was done. However, he did have the biggest bruise I have ever seen in my life. Following this painful incident, we rarely ventured to see the horses. I haven't seen David for 30 years, so I have no idea whether he suffered any long-lasting psychological consequences as a result of the kick, but I must profess to developing a mild phobia of horses. Not for a moment have I ever felt that the horse that kicked David was in any way responsible for the incident. Even so, for many years I was very reluctant to touch or pat a horse and I would certainly not entertain the idea of ever riding one. However, as I entered my thirties, I started to feel a little more comfortable in the company of horses. My re-adjustment began on a trip to Norway in the summer of 1993. I was on a two-week holiday with my family to Bergen. We were staying with some close friends and their house backed on to a stables and riding school. My daughter, Laura, then six years old, was very eager to see the horses and I was very happy to accompany her, which in itself surprised me given my history. The holiday couldn't have been better timed because I had been enduring particularly high levels of stress in my mental health work. Even though the vast majority of people suffering from prolonged problems with their mental health present no threat to others, I had been working very closely with an extremely aggressive client who was potentially quite dangerous. I was exhausted and looking forward to de-stressing in the peace and tranquillity of the Norwegian fjords. The idea of visiting the stables had a similar appeal. Spending time with some horses, at a safe distance, seemed another possible route to relaxation. I was not disappointed.

Our Norwegian hosts quickly arranged for us to visit the stables. The riding school's facilities were quite extensive with a large indoor arena. Laura was keen to go into the stables and we were welcomed in by the stable hands who, like most Norwegians, spoke very good English. As we entered we were confronted with a large Irish stallion named O'Mally. He was being groomed by one of the hands who began telling us what a wonderfully gentle horse he was. This was as close as I had been to a horse for over 25 years. There was no fence between us either. O'Mally was out of his stall and I was standing calmly just inches from his side but I felt no fear. I began to stroke him and soon sensed the heat of his body. His warmth was both calming and reassuring. Without thinking I put my head on his flank. As I did so he turned towards me and looked me over. I didn't move and he gently nodded. All too quickly the time arrived for us to return to our hosts. I gave O'Mally a light hug and told him I would visit him again during our stay. He certainly did help me relax. His reassuring warmth gave me what I can only describe as a glow of contentment. He also gave me back the confidence and trust to enjoy the company of horses again. It was a good feeling, but I was still unsure whether I felt ready to ride yet. The time would come, but I would have to wait just a little longer before I met the person who would eventually inspire me to overcome my fears.

My meeting with O'Mally proved to be my first experience of the therapeutic benefits that can be derived from horses. However, equestrian or riding therapy as a means of enabling personal development, education and healing for people with a range of disabilities and special needs has been well established throughout the world for a number of years. Probably the UK's most well known organisation promoting this activity is the Riding for the Disabled Association or RDA. The Association grew out of what had been known as the Advisory Council on Riding for the Disabled. The Advisory Council was established in 1965 when nine riding therapy groups came together to form a national organisation. In 1969 this became the Riding for the Disabled Association, by which time it consisted of over 80 groups. By 1998 this had grown to some 700 hundred groups with more than 23,000 riders and drivers.

The RDA offers the chance to ride to any person with disabilities who might benefit from doing so in terms of their general health and well-being. They have experience of supporting people with a range of disabilities, including those with cerebral palsy, spina bifida, multiple sclerosis, muscular dystrophy, multiple injuries, those without limbs (including people disabled by thalidomide), as well as those with learning difficulties and sensory impairments, both visual and audial. People who are over 14 years of age and physically prevented from riding are offered the opportunity to learn to drive a pony- or donkey-drawn vehicle.

The RDA has affiliated groups in many parts of the world. These have been established in Europe, North and South America, Africa, The West Indies, Singapore and, more recently, in Russia and Japan. I was very privileged to meet the International Liaison Officer for the Association, Sister Chiara Hatton-Hall. As a Fellow of the British Horse Society, and an extremely experienced rider and trainer, she became involved with RDA in the early 1980s. Sister Chiara is a Franciscan nun and has been a member of her order since 1974. She began working with the Association when she completed her noviciate. Her main involvement over the years has been on the development of training skills for the volunteer instructors who work with the disabled riders. In 1991 her role expanded to include working with the affiliated groups abroad. She also works as an examiner for the Association and has only recently retired from performing a similar role for the British Horse Society over the last 40 years.

Sister Chiara offered some personal insights on the benefits of riding for people with disabilities, drawn from her many years of experience of the human-horse bond. Her thoughts did much to illuminate my appreciation of the more academic appraisals of this type of therapy. My personal knowledge of horses is, as I have described, somewhat limited, but the therapeutic effects of equestrian therapy appear to fall into three overlapping categories: personal development, education and physical benefits.

Personal Development

Sister Chiara believes that there is a particular magic about horses and that most people who become involved with them develop an enduring bond and love. She thinks this is because we need an affinity with

nature and the horse is a highly sensitive and responsive ambassador. If a horse is treated gently and with
respect he will respond accordingly. The horse offers a
consistency of response and an uncomplicated relationship. For many this is a rare commodity in human
relationships and, for those who may have been isolated because of their disability, the feeling of acceptance can do much for their feelings of self-worth. The
horse is uncritical and non-judgmental and has no
sense or knowledge of past failures. Many companion
animals share these attributes but the horse may prof-
fer a particularly potent therapeutic cocktail. In an
earlier chapter, I identified enhanced levels of social
support, control and environmental change as the
three major factors that influence the level of therapeutic affect derived from companion animals. If we
consider each of these briefly in relation to horses, it
may illustrate why many believe that horses are a
primary purveyor of animal assisted therapy.

The social support derived from horses may be qual-
itatively different from many other companion
animals. Due to the obvious impracticalities, we are
unable to directly share living environments, but the
comforting, non-judgmental and consistent nature of
their company is undiminished. Sister Chiara described to me the solace she derived from her pony as
a young girl. If she was upset about anything, she
would run to the stable, often bursting into tears and
then burying her head in the pony's mane. I'm sure a
pony listens as well as any dog or cat. They also require
an enormous amount of grooming and looking after.
For many, the horse provides a much needed opportunity to care for and nurture another living creature.

It is in relation to control and environmental change
that I feel horses particularly excel. The riding of a

horse demands that the rider controls the relationship. Since the horse is the biggest and most powerful of companion animals, the sense of achievement and feelings of enhanced self-worth and self-esteem may be magnified accordingly. For those people with disabilities or special needs who are unable or prevented from controlling many aspects of their lives, the feelings of accomplishment must be profound.

For most people the riding of a horse necessitates a complete change of environment. Stables and riding schools with their very particular sights, sounds and smells form the setting for the most significant environmental change of all – sitting on a horse. This position – sitting on an animal's back – is restricted to very few companion creatures. When riding, one moves through the environment in a different mode and shares the animal's rhythm. For many people with disabilities who are confined to wheelchairs, sitting astride a horse or pony may also be their first experience of literally looking down on people. A person's social surroundings are also altered, by meeting new people, whether instructors, support staff or fellow riders. It is not unusual for people with disabilities to excel at riding, and this may lead to particularly significant changes to their social environment, since a person with disabilities may find themselves in competition with fully-able people. In this respect the horse can act as a very effective equaliser. There is also another activity where people with and without disabilities sometimes come together on equal terms – the musical ride. This activity is also known as kur or dressage to music. Sister Chiara expressed, if you will excuse the pun, an unbridled enthusiasm for formation riding to music. RDA teams have performed in competition at Ascot, Cheltenham and Kempton Park race courses, and at the Wembley International Horse Show and Windsor.

Education

Experience has demonstrated that those who develop a close affinity to horses, often show greater motivation to learn. An equine environment can provide a more stimulating climate for education which can be more effective in holding an individual's attention. This complementary classroom offers an alternative approach for children and adults with or without disabilities who have not flourished in traditional teaching settings. On a simple level, time with or riding horses has been used as an incentive to encourage people to attend to their classroom studies. However, horse-motivated education can offer much more. Stables and riding schools present settings littered with opportunities for applied learning. Caring for a horse involves counting, measuring and weighing. How many bales of hay are needed? What weights and quantities of feeds are required for each horse? How many stalls are still to be cleaned out? I have quite recently become aware that letters are used as marker points when learning to ride. Also integral to the process is familiarising yourself with line, shape and distance. These applied uses of mathematics and letters offer a more tangible and less abstract learning situation.

Learning to groom and care for horses is also a specialised area in its own right. Sister Chiara recounted just one example amongst many that demonstrates the potential of this area of education. A young disabled boy began riding at the age of ten with the RDA at the

Downing Centre. He became a very able rider and led a musical ride with great success. At the age of 15 he left the centre and went to work at a racing yard. He is now permanently employed at the yard where he grooms and rides out horses. Learning to groom horses has also encouraged many young people to pay more attention to their own personal care and presentation.

Physical benefits – hippotherapy and vaulting

Hippotherapy refers to the physical benefits derived from riding, where the accommodation of the swinging motions of the horse by the rider stimulates and exercises parts of the body. The natural movement of a horse produces movements in the rider similar to the human walk. Horse riding offers moving physiotherapy and encourages body symmetry. Sister Chiara explained to me that riding a horse can produce up to one thousand random body movements in as little as ten minutes, whereas such a workout would take up to three months of normal physiotherapy. The horse has another advantage in that his natural warmth aids relaxation. The success of hippotherapy has been demonstrated by X-ray and electromyography. Vaulting, the assisted performance of gymnastics on horseback, has also been shown to offer both physical and psychological benefits in terms of balance, co-ordination and confidence.

Supportive Research

The findings of several research studies have clearly indicated the effectiveness of this form of animal assisted therapy. A study conducted at my local hospital, Queen Mary's, London, was so unequivocal in its findings that it led to an almost immediate expansion of the small-scale programme that had been the focus of the research. The study in 1969 evaluated the effects of riding on six people, three who were physically disabled and three who were diagnosed with learning disabilities. After only a few weeks the group showed significant improvements in behaviour, language, communication and physical functioning.

In an American study Natalie Bieber evaluated the effects of a five-week equestrian therapy programme on a group of 42 children aged 6–17 who suffered from a range of disabilities, including spina bifida and cerebral palsy. The study involved the children in riding for one day on a horse or in a pony cart, and two days in the classroom using horse and horse-related material as an incentive for learning. Staff who assessed the children found that all but four benefited significantly from their involvement in terms of communication and motivation. The programme also appeared to stimulate the children physically, socially and intellectually.

In 1975 research was undertaken to assess horse riding as a risk exercise, and as a means to increase self-confidence, courage and motivation. One hundred and two physically disabled children at therapeutic riding centres in England, Ireland, Wales, Canada and the United States were studied. The study found a high level of improvement in mobility, motivation and courage. The morale of many of the children was also greatly enhanced.

A seven-year study was conducted in Washington DC into the benefits of a riding therapy programme for physically disabled people and those with learning difficulties. The programme was evaluated on a yearly basis, using input from teachers, parents and the students. Analysis of the evaluation revealed some startling results. As a group, it was gauged that there was an average gain in physical movement that ranged from 7 to 31 per cent. Eighty per cent of the children were found to have improved language skills, with the average gain being between 9 and 29 per cent. Average increases of 6 to 19 per cent were found in emotional control, social awareness, peer relations and self-awareness. Seventy per cent showed notable improvement in work skills, with an average gain of 17 per cent. Eighty-seven per cent of the parents commented upon their child's improved self-confidence and there was a 52 per cent decrease in the number of negative statements made by the children. The teachers' overall evaluation of the effectiveness of the programme was 'very good' or 'excellent'.

Another piece of American research was conducted by Ruth Dismuke into the effects of therapeutic horse riding on children with language disorders. Thirty children whose ages ranged from six to ten years old were classified as moderately to severely language disordered. The children were matched for age, type and degree of language disorder and were randomly assigned to the experimental or control group. All the children received language therapy for three one-hour sessions per week for 12 weeks. The control group received therapy in a state school therapy setting. The experimental group was treated in a structured horsemanship programme in which speech and language specialists were also professional riding instructors. Independent testers who were unaware of the children's group placement evaluated their improvement through tape-recorded conversations.

The results indicated that the horsemanship programme facilitated the language therapy. Although
both groups demonstrated more complex sentence
structure following therapy, the experimental group
exhibited the ability to use their language more efficiently and appropriately. The study concluded that
the use of riding appeared to have enabled an enhanced development of language skills. In addition,
significant gains in muscle strength, co-ordination and
self-esteem were noted.

The Fortune Centre of Riding Therapy

Riding unquestionably provides a holistic therapeutic
approach that addresses psychological, physiological
and emotional aspects of human health. There are
many centres that offer special needs riding programmes throughout the world. One such centre, the
Fortune Centre of Riding Therapy in Dorset, southwest England, has the enthusiastic backing of the
champion jockey Lanfranco Dettori. The Fortune
Centre has developed a range of programmes with
proven benefits for young people with special needs,
including those with learning disabilities and behavioural and emotional difficulties. The centre provides
full-time residential education for young adults
between the ages of 16 and 25 who seem to be motivated in a horse environment. The Further Education
Through Horsemastership (FETH) course, which
normally lasts two years, teaches life and social skills
using a horse-based extended curriculum. In essence,
the unique qualities the horse provides are matched to
the willingness and motivation to learn of the individual. Students come from all over the world but are

mainly from Great Britain. They are involved from the beginning of the course in developing their own learning plans and are expected to acknowledge their difficulties in order to change and move forward. Students are encouraged to achieve their personal best. For some this will mean passing formal qualifications.

Longer-stay residential homes are also provided, and these are available to ex-FETH students who have more enduring special needs. 'Ostlers', as these residents are known, are not able to manage daily living without on-going help and support. However, an Ostler home enables them to live independently of their families with a like minded peer group, involved in horse-based occupations. Ostlers spend five days a week in supervised work with the horses at one of the Fortune Centre's farm locations, but are encouraged to mix with the local community and enjoy local leisure pursuits. Residents of the Ostler homes can remain until they are 55 years old. However, some do move on as they develop or their needs change.

The Fortune Centre also offers weekly riding therapy sessions to local children with special needs from 2 to 12. In addition, professionals employed by the centre – teachers, riders and care workers qualified in their own disciplines – have the chance to undergo in-service training as riding therapists. There are also a very limited number of opportunities each year for students on their last year of relevant graduate courses to undertake placements at the centre. The Centre also provides a support and information forum for students' families and potential employers, plus it acts as a source of information for the general public concerning the meaning and practical function of riding therapy in order that more people can benefit from the help available.

Green Chimneys, New York

Horses play a very significant role in the Green Chimneys' programme in New York. This highly regarded project offers residential care to young people from 6 to 21 with a range of special needs, mostly from New York and the surrounding area. Many of the children have histories of neglect, or sexual, physical or emotional abuse. A good number also have deficits in their education as a result. Over 100 youngsters share the 150-acre site near Brewster in New York, with a range of domestic, barnyard and rare animals. The staff have pioneered the use of human–animal interaction as the foundation for their therapy work with these young people for the last 50 years. Samuel B. Ross Jr, the founder of the project, believes that the children 'come defeated because they have failed in those things by which children get judged. They must learn that there has never been an animal which asked a child his achievement test score.'

The guiding philosophy at Green Chimneys is that, through their work with the animals, children begin to learn responsibility, and by nurturi g the animals they may begin to nurture themselves. As Samuel B. Ross Jr notes, 'Very often the children feel depressed, with-drawn and unwanted. They need to feel a sense of con-nection, a personal bond with another living thing. For many a connection with an adult or peer is threaten-ing. In many of these cases the animal is the logical answer. The lessons learned from the animal become a stepping stone for a human connection. Nurturing an animal and receiving back unconditional attention and love re-establishes the worth of the child. It encour-ages the child to risk the human connection.'

Children care for a range of animals, including

injured turkeys, geese, owls and falcons at a disabled wildlife rehabilitation centre within the project. They also work with horses, ponies and farm livestock. Other domesticated animals cared for by the children include dogs, cats and rabbits. Every child at Green Chimneys adopts a horse. They may ride the horse whenever the programme allows. The care of the horse becomes the child's responsibility, with appropriate supervision if the animal is injured or becomes sick. Children reluctant to ride are still involved in caring for a horse and undertake equine studies. Therapeutic riding instructors provide classes for both the residents of Green Chimneys and for visiting mentally and physically disabled young people. The resident children are encouraged to act as aids to the instructors, helping their disabled visitors mount the horses. A job experience scheme offers some older residents the opportunity to be trained in working with riders with disabling conditions, either as a therapeutic riding leader, or sidewalker. Residents learn to care for other people and become more 'service-oriented' as a result of this experience.

Samuel B. Ross Jr cites the following letter, sent from one child to another following the death of his adopted horse, as an illustration of what the horse care programme means:

Dear RD,

I am sorry about Jagger. He was a special horse. We all loved him. If there is anything I can do to make you feel better, I will do it because I know how you feel. I am upset myself. I know you will miss him, for I will miss him too.

Your Friend
MP

'Jagger's death was an emotional time for many of the children. Many had family or friends who had died or were killed. It gave them a chance to mourn old friends and ask many questions, along with coming together to support one another. For children who might have trouble expressing themselves in verbal therapy in an office, animal assisted therapy becomes a vital link in a child's treatment.'

Against the Odds

To illustrate the potential horses have to enable people to come to terms with, and in many cases overcome, their disabilities and difficulties, I would like to share the stories of some runners and riders who serve as an inspiration to us all.

Probably one of the most celebrated examples of a rider overcoming their disabilities is that of Liz Hartel. She contracted polio in the 1940s and although confined to a wheelchair, she went on to win a silver medal for Denmark in the 1952 Olympic games. Her achievement did much to encourage the establishment of the Riding for the Disabled Association.

When I visited Val, the PAT dog volunteer, in the north of England, I noticed an enormous amount of trophies on display. I enquired what these were for and she explained that her involvement with animal assisted therapy began through her horse riding. She had been a successful amateur horsewoman and realised that riding could serve a therapeutic purpose. She taught her severely disabled husband, Vic, to ride her horse, Flare, which necessitated her inventing stirrups on a swivel to accommodate his artificial legs. She also perfected some stump socks which didn't wrinkle to prevent him from developing sores. Horse riding was a great equaliser for Vic and did much to enhance his self-esteem. Val explained to me that Flare could be a little naughty but the moment Vic was on her back she behaved perfectly. She was so gentle that Val described her movements as though she was 'walking on diamonds'. If Vic slipped or suddenly moved Flare would stop and look round to see if he was OK before walking on.

Flare was also a great comfort to Val during her many years as a nurse. At one point in her career she was a Sister of Midwifery in a large maternity unit. She supervised 11 delivery suites and even though this work could be most rewarding, it could also be very draining and emotionally demanding, especially when life-threatening complications arose. On several occasions following a busy late shift, a very exhausted Val would go to the stables to see Flare. The horse seemed to exude an uncomplicated warmth, not asking anything of Val and seemingly waiting to offer some support. Val would sit on the stable floor and Flare would rest her head on Val's shoulder, who was exceedingly grateful for her concern and company.

Val seemed to me to be a woman who lived her life by the adage that actions speak louder than words. When she described the following events, any lingering doubts I may have had about this soon faded. She had read in the local newspaper of young girl, Debbie, who lived close by and who, due to cancer, had recently had her leg amputated at mid thigh. From the detail given in the article, Val was aware that the girl, as one might expect, was having great difficulty coming to terms with facing the future with one leg. On reading the article, Val decided to ride over on Flare and visit her. Val knocked on the door of Debbie's house which was answered by her mother. Val introduced herself, spoke of Vic's disability, and asked to see Debbie. Her mother was quite distraught and explained that her daughter wasn't prepared to face anyone just yet. Val, as you may have guessed, rarely takes no for an answer. She explained that she was keen for Debbie to see Flare and asked if she would just come to the door because she might like to get on the horse. Val can be quite persuasive and Debbie eventually appeared. Val invited her to just sit on the horse. At first she declined the offer but with a little more prompting and persuasion she agreed, and got on the horse. Debbie seemed to enjoy it, but was reluctant to do anything more than sit.

Val was not discouraged and so a couple of weeks later she asked Vic to get on Flare and ride with her to see Debbie. Debbie came to the door and Val introduced her to Vic, but not in the usual way. She lifted his trousers and showed Debbie his artificial legs and explained that he had ridden from their home while she had walked. This proved to be a turning point for Debbie. She began to ride the horse on a regular basis and Flare was as supportive as possible, taking great care with her new companion. As soon as Debbie was confident and competent enough, Val allowed her to take Flare out to ride on her own. Val and Flare did much to enable Debbie to come to terms with her disability and riding was just a beginning. In subsequent years she took up skiing and has competed in the Paralympics.

When I met with Sister Chiara she had recently returned from a lecture tour of Japan. She explained that in recent years riding for the disabled had become increasingly popular there. In 1998 a Japanese affiliate to the Riding for the Disabled Association had been established and she was very pleased that they were now part of the worldwide network. Sister Chiara told me of Madoka Hoshino, a 25-year-old woman who suffers from cerebral palsy, who had been chosen as Japan's first participant in an international riding competition held in Gloucestershire. Madoka has spent most of her life in a wheelchair. Horse riding was recommended to her because she had trouble controlling her legs and it was thought that it would be good exercise. When she first began riding she found it difficult to sit up straight but she overcame this and learnt to ride unaided. By any assessment, Madoka's riding achievements are most impressive. However, what is particularly startling is that at the time she was invited to compete in the international competition, she had only been riding for two and a half years!

In the summer of 1997 I travelled to Newton Abbot in Devon for the annual conference of the Society for Companion Animal Studies (SCAS). The two days of the conference were equally divided between drafting guidelines regarding the welfare of animals (non-human and human) involved in animal assisted therapy programmes, and visiting the Camomile Centre, which was located just outside Bovey Tracy, in 11 acres of open land on the edge of Dartmoor, and offered animal related activity therapy to children and adults with a range of special needs. The centre, under the stewardship of its founder, Helen Cottington, had been modelled on very similar lines to the Fortune Centre but also included a number of other animals in addition to horses – donkeys, sheep, hens, dogs, rabbits and guinea pigs.

On the first day of the conference when we were

formulating guidelines, we divided into groups of six or seven people to focus on more specific aspects of animal welfare. I was immediately struck by the presence of one of my group members, Carrie, who seemed a good deal younger than the other participants. Her articulate, sensitive and perceptive comments belied her years and it soon became clear that her insights were born from first-hand experience. It hadn't been immediately apparent but, as the morning wore on, I realised that one side of Carrie's body was in some way physically impaired. I chatted to her at the first coffee break and she told me a little of her background. She was 18 and a student at the Camomile Centre. She had been born with haemoplegia – a fairly extensive paralysis to the right side of her body. She spent most of her time at the centre working with the horses and had become particularly attached to one of them, Merlin, a 12-year-old blue roan Connemara cross. Carrie confided in me that she was a little nervous because she was going to give a riding display the following day when the conference delegates made their visit to the centre. I tried to offer some reassurance by telling her that as far as I was concerned, anybody who rode had my respect because I broke into a cold sweat just at the thought of it. She smiled at this but as we spoke a little more I became increasingly aware of just how important her ride the next day was to her. Unless one is blessed with the gift of pre-cognition, we human animals are never aware of the poignancy of certain moments and episodes until after the fact. With this firmly in mind, I will now share some insights, only gleaned later, about Carrie's past. I hope this will enable me to do justice to the significance of the events that unfolded the following day.

The difficulties Carrie experienced as a child due to

the partial paralysis of the right-hand side of her body, were compounded because she was born right-handed. As we are all aware, using our 'other' hand is awkward and restricting. For Carrie this resulted in her being considered a slow learner in her first years at school. As a consequence of this, several of her teachers during this period took the easy route and put little effort into her education. Clearly, she did not have the most auspicious start in life. Sometimes I wonder at the seemingly unending and unexplainable unfairness of life, and in Carrie's case there was still plenty more to come. She also had to undergo surgery on her tendons. This left her with one leg, her left, slightly shorter than the other.

By the age of seven Carrie had begun to develop an interest in horses. Her parents discussed the possibility of her riding with her hospital consultant during a check-up. The doctor informed them that Carrie would never be able to ride due to her physical condition. To one so young, his comments could have been enough to resign her to a life of dependency, inactivity and regret, but not for Carrie. His words served only to fire her determination and she very quickly set about proving him wrong.

With the support of a local RDA group Carrie was riding within a year. She stayed with the group for two years or so and then moved on to a riding school at Honeysuckle Farm where she further developed her skills and quickly learnt to canter and jump. It was here that she met Yanna, a 12-year-old New Forest pony who was to become a very special part of her life. They developed an almost instant rapport and nearly always rode together at the farm. Carrie felt they looked after each other. She would groom and care for Yanna, who would be there for Carrie when she needed a friend. If she had an argument with her parents she would go to Yanna and sit in her stable and talk to her. Yanna would put her head over Carrie's shoulder which gave her some close comfort and just listened. She never answered back and seemed to understand when others couldn't. Their friendship continued to blossom and they rode together for nearly four years.

Just after Christmas 1990 Yanna developed colic, a very painful and life-threatening condition. She became very sick and on New Year's Eve Yanna was put down. Carrie was discouraged from being with her at the end, but Yanna's owners gave her the horse's brow band and name tag from her bridle which she has kept to this day. Carrie was devastated by her loss and found it too painful to return to Honeysuckle Farm. She didn't ride for over two years following Yanna's death.

However, Carrie began riding again when she started attending the Camomile Centre. Initially this proved to be none too easy. She lost confidence after falling off one of the horses at the centre when she attempted a jump but with the help of Moley, a rather distinguished 30-year-old blue roan Dartmoor pony, she was soon jumping again and getting back to her old self. She was very hesitant about getting too attached to any of the ponies but before long Carrie began to become very close to Merlin. It took a little time for them to establish a mutual trust but the bond that eventually developed between them became very strong indeed. Carrie felt that she could communicate with Merlin, and after witnessing Monty Roberts demonstrate his ability to understand horse language, her feelings were confirmed. She had inadvertently adopted some of the same techniques as him.

She had been attending the Camomile Centre for two years when concerns began to arise about Moley. He was now well into his thirties and his health was deteriorating. He was very much loved by all those who attended the centre and Helen Cottington was very keen to do what was best for him. He had offered support through riding to many students. He was also the very first pony that my daughter Laura had ridden. I had only met him briefly myself, but he seemed a very gentle creature.

After consulting the local vet, it was decided, very sadly, that Moley would be put down. Helen and her staff tried to prepare the students and themselves as best they could. Everyone decided where they wanted to be at the allotted time. Helen had to leave the centre because it was too painful for her. Carrie wanted to be with Moley until the end. On his last morning she led him up the lane to the centre for a last feed on some grass. When the final moment came, she was at his side. Carrie felt she owed Moley a great debt of gratitude because seeing him painlessly put down enabled her to accept that this is often the best way to prevent suffering. She now realised that for Moley and Yanna, the right decision had been made.

The conference delegates' visit to the Camomile Centre began with a tour of the facilities, which was followed by some very informative talks given by the staff on animal related activity therapy. Even though it was June the wind from the moor was gusting and the clouds were beginning to darken as I spoke to Carrie just prior to her leading Merlin out for the display. I wished her good luck and took my place with the rest of the delegates at the riding arena. There were about 50 people gathered around the perimeter fence, and, right on cue, minutes before the horse riding event was about to begin, the rain began to fall. It wasn't particularly heavy but, together with the strong wind, conditions were not the most comfortable for riders, horses or spectators. The riders in the demonstration had a range of special needs and included both children and adults. They demonstrated their control and confidence as they performed a number of riding exercises. As I'm sure you've realised by now, I am very much a novice in the equestrian field but the way the students worked and interacted with the horses seemed to impress even those with the most expert eye.

Carrie was last to ride and as she mounted Merlin I began to feel a little tension rising from within. I so wanted her to do well. She made several circuits of the arena and then rode him over and through a number of obstacles. She then dismounted and removed Merlin's saddle. For moment I thought the ride was over but within a few minutes Merlin, with no lead rope, was following Carrie around the arena in different directions, responding almost instantly to her every prompt and request. In my mind I had no doubt that she was communicating with Merlin in a language that he clearly understood. After several minutes it appeared that Carrie had told Merlin to stop and stay where he was. She then walked about 15 metres ahead of him and then stopped with her back to him. On her command Merlin walked up to her and then rested his head on her right shoulder. Spontaneous applause followed but Carrie was not quite done. For a grand finale she remounted Merlin and rode him bareback for several circuits of the arena. Her final dismount was greeted by the rousing reception it justly deserved. I had very large lump in my throat by this point. It was the most moving display and when I managed to work my way through the congratulatory throng to offer my praise, I could hardly speak. However I did manage to express my admiration and thank her, not only for the most wonderful experience, but for her inspiration, because after witnessing her demonstration, and after nearly 30 years of fear, I at last felt ready to ride. Carrie and I hugged and then agreed that once I had taken a few lessons, I would join her for a ride at the centre. The euphoria generated by the riding was soon tempered by the news that due to a financial shortfall the centre was very likely to have to cease its activities. Sterling efforts were made but that unexplainable unfairness of life struck again and the Camomile Centre closed at the beginning of 1998. The centre had offered hope and opportunity, and met the needs of so many. All is not lost, however, because at some point in the future Helen Cottington hopes to resurrect the centre or something very similar.

Once again though, Carrie had to endure the loss of a close friend. Merlin, as well as the other animals, had to be found a new home. At the time of writing Carrie had not see Merlin since the centre's closure. She explained to me that she thought it was very important for her to come to terms with her feelings about their separation before seeing him because he could be so receptive to her emotions. She wanted to be happy for Merlin when she saw him and the last thing she desired was to disturb his new life. For a period of several months following the closure of the centre, Carrie rode far less frequently and then stopped altogether. Seemingly as a consequence, she began to experience problems with her lower back. She had always found riding kept her body toned and loose. However, getting back on a horse has been the story of her life and she has now begun riding once again. Her enthusiasm has returned and she needs no convincing of the benefits.

Carrie believes that animals, and horses in particular, are more open than humans and readily trust their instincts. By learning to ride, she became aware that her disability was not a bar to achieving her fullest potential. As a result, she is a most determined and motivated woman. To date, she has advanced vocational qualifications in animal care and the developmental care of humans. After a recent period of study she is presently working as a volunteer at a local facility for adults with learning difficulties. In our most recent conversation Carrie informed me that she had just been to the hospital for a check-up and to her to surprise she was seen by the same consultant who nearly 15 years earlier had spoken those fateful words. Well she certainly proved him wrong. The now quite elderly doctor, greeted her saying, 'I hear you're a horse rider now' and gave her a very cheeky smile which left Carrie in little doubt that she had been the victim of a rather successful ploy. From what she told me, I think that he was almost as pleased as she was that the subterfuge had worked.

As for me, well I've started to ride and even though my first lesson proved to be quite nerve-racking, I'm very pleased I took the plunge. I went with Laura to the Ealing Riding School in west London. I'm not familiar with the sights and sounds of a busy riding school and even less with the smells. A blacksmith was shoeing a number of horses in the stable courtyard so a pungent aroma of burning horse hair combined with the heady mixture of horses, hay and manure greeted us on our arrival. This was indeed a different environment for me. I quickly became aware again of the natural affinity between children and animals. Laura rode a rather beautiful young pony, Crystal. They seemed to strike up an almost immediate understanding and Laura readily accustomed herself to the rhythm of the ride. I sat astride an older grey called Paddy who was very gentle with me but I struggled to get into the trot. I was led by a 12-year-old girl on my first lesson, who did all she could to reassure her charge. However, after a little time had passed I began to feel more relaxed. This long awaited experience was all that I hoped and I felt rather elated. The closeness, sharing and trust between horse and rider enables a tangible bonding with nature often missing in our lives. The therapeutic and restorative benefits of riding are clear and my subsequent lessons have certainly proved an effective means of reducing my day-to-day tensions. I certainly intend to ride whenever I can and hopefully, before too long, with Carrie, to whom I owe a rather large debt of gratitude.

CHAPTER 7

IN THE CARE
OF CATS

My relationship with horses, and more generally my work as a psychologist, have left me in no doubt that difficult childhood experiences can affect us throughout our lives. To overcome these unresolved fears or feelings, we may decide to confront them with the support of a counsellor or therapist and, if possible, put them to rest. However, at other times, we are inspired by the deeds of others – as I was by Carrie – to face our past head on and just 'go for it'. Every now and then however, if we are lucky, experience gently persuades us that our fears are ungrounded. I tender these words as a means of offering my personal perspective on our feline friends. As a six-year-old child I was bitten by a rather aggressive tom cat and, even though I held back from judging all cats by this maverick's behaviour, I must profess to spending many years feeling quite indifferent to their existence. Now and again I met an affectionate puss who I warmed to slightly, but I never sought their company as I would a dog and a bond was never established. That was until I met Sooty.

I made Sooty's acquaintance through my future wife Patsy, because he was the very close companion of her then eight-year-old daughter, Olivia. Sooty was a largish seven-year-old black and white neutered tom. He was a very special friend to Olivia, because at that time in her life friends were pretty hard to come by. As a young baby Olivia, or Olly as she is more usually known, developed infantile eczema, the disfiguring skin disease. She was covered from almost head to foot in an angry red and often weeping rash. It is a very uncomfortable and, at times, painful condition. The sores can become almost too itchy to bear. As a consequence, young children have to be bandaged a good deal of the time, especially at night, to stop them scratching the affected patches in an attempt to prevent further inflammation and possible infection. The condition is not contagious, but most unsightly, and in the playground it can prove to be more than enough to leave a child isolated and friendless. Olly also suffered from a number of allergies, including a sensitivity to fur, so some of her early years were pretty miserable. She had the occasional friend, but spent much of her first six or seven years alone, except for Sooty. He was a cat with a number of unusual qualities and quite remarkable abilities. For a start, and this may have crossed your mind already, there was the fact that he was hypo-allergenic. Olly used to spend many hours in his company stroking and playing with him, and he also used to sleep on her bed most nights, but he never once caused Olly to have an allergic reaction.

Sooty had a particular passion for hot chocolate and his method of consumption enthralled me on a number of occasions. I remember vividly the first time I witnessed him feast on the residue of Olly's bedtime drink. One cold winter's evening, quite soon after I'd met Patsy, we were sitting with Olly just prior to her going to bed. She asked for a mug of hot chocolate as she was feeling the cold a little. This may well have been a ploy to delay bedtime because she had declined an earlier offer. However, Patsy was more than happy to respond to her request because Olly was rather a fussy and picky eater, so every opportunity to get something warm and nourishing inside of her was gratefully seized. Olly slowly sipped at her warming drink and seemed to pause over every mouthful. I was soon aware that Sooty had joined our number and was at Olly's side with his eyes firmly focused on the mug she was holding. She eventually finished the drink and said 'it's all gone but the dregs'. Patsy responded by enquiring whether she had left any for Sooty. 'Oh, of course Mummy', was the reply and she then placed mug in front of the rather pleased looking cat. Firstly he sniffed at it and then began scooping out the cooled chocolate residue at the bottom of the mug with his paw, which he then eagerly transferred to his mouth. He continued extracting the sticky substance this way until it was all gone. Olly explained that he always did this with hot chocolate but she had never seen him do it with any other food or drink.

Patsy lived in the basement flat of a converted house and Sooty had two means of entry, the living-room window or the front door. If the window was closed and he could see we were in the room he would scratch or lightly tap on the glass with his claws. We would open the window and in he would come. However, his scratching or tapping on the front door could not be seen or heard. This didn't seem to be a problem for him because, if he was sitting on the mat outside the front door, one of our kindly neighbours would see him there and knock on his behalf and we would let him in, or so I thought. It wasn't until some time later that, purely by chance, I saw Sooty balancing on his hind legs with one of his paws supporting his weight, while using the other to lift and drop the door knocker as he 'knocked' on the door. As I watched him with my mouth gaping wide Patsy opened the door and he marched in. I excitedly told her of my discovery, but she didn't even raise an eyebrow as she said, 'Oh he's always done that'.

I grew very fond of Sooty as the months went by. It was apparent that he meant the world to Olly and her feelings seemed to be wholeheartedly reciprocated.

Unfortunately, six months or so before I met him, he had begun to develop growths in his ears. Initially they were not considered to be serious but his condition worsened and he underwent an operation to remove the tumours. He made a speedy recovery and all seemed well. Sadly, within a year of their removal they returned and their malignancy was confirmed. I was working on a construction site the summer Sooty was put to sleep. I received the news when I was called to the site office for a phone call from Patsy. The inevitable can still shock and when I rejoined my work-mates I was a little pale. They were keen to know why I had been called away. When I shared the news with them and gave a little background detail, I was quite taken back by their reaction. Several of these hard-ened, and at times quite intimidating, individuals were close to tears.

Olly was inconsolable. No words of comfort could ease her pain and she cried herself to sleep for many nights. She lost Sooty two weeks before her ninth birthday. He had been in the family since she was 18 months old. Eventually her pain began to ease but occasional tears were shed up until her early teenage years. Patsy, too, suffered but an unexpected arrival did much to help the healing process begin. It must have been about seven or eight weeks after Sooty's death that on one sunny October morning we noticed some rustling in the bushes outside the living room window. As we couldn't see what was causing the leaves to shake we thought no more of it. Patsy and I left the room and got on with some household chores but Olly continued to watch through the window. After five minutes or so Olly cried out 'Mummy, Mummy come quickly'. Patsy's first reaction was one of concern because it sounded as though Olly had had an accident. We rushed into the room, however, only to find her very gently stroking a most beautiful jet-black kitten who couldn't have been more than five weeks old. He had apparently popped out of the bush and affected an unforced entry through the open window. Patsy inspected this most welcome intruder and for the first time ever I held a kitten while she tried to determine its sex. She was pretty sure he was a tom. He had no collar and we thought he might be a stray. Of course Olly's immediate wish was to keep him but we knew we must try to find his home before we could entertain the idea.

We called on our neighbours but nobody was aware of any missing kittens. Olly named him Lucky on a temporary basis, not only on account of his colour but also because she felt that she had been very lucky that he had come to her home when she was so sad about Sooty. We made every effort during the following week to find his home but to no avail, so we decided to adopt this most affectionate and playful creature. Over the next few weeks he quickly established himself as very much a part of the family. I immediately fell in love with him and on one occasion when he was quite upset because he had become trapped at the bottom of a tea chest, I rescued him and he licked my hand. I was a little surprised at this because I'd never been licked by a cat before and was unaware, even in a cat so young, that his tongue would be so rough. Lucky did much to help Olly begin to come to terms with Sooty's loss. Even though Lucky's presence occasionally irritated her skin, they, too, became close companions. We were all very pleased to have him. Patsy and I never did find out where he had come from but Olly thought she knew the answer to the mystery. When she informed us of this we were very keen to hear what she had found out. In unison we asked, 'So where is he from?' Her response is now family folklore, 'God sent him because he knew how upset I was about Sooty.' I'm not a religious man but her answer was good enough for me.

Due to cats' independent nature, I was a little surprised to hear of a feline equivalent to PAT (Pets as Therapy) dogs. PAT cats was launched in November 1997 at the Supreme Cat Show at the National Exhibition Centre in Birmingham. The very first cats enrolled into the programme were three Persians, Toby, Chiffon and Woody. They regularly visited a psychiatric hospital and a school for children with special needs. Any cat who has been with their owner for six months and over can apply to become a PAT cat. Every applicant is thoroughly examined by a vet and undergoes a temperament test. The test is conducted to check if a cat is sociable, friendly and not nervous or aggressive. A PAT cat must be calm and gentle when being groomed or stroked. The vet conducting the examination also checks a cat's reaction to sudden noise.

At the time of writing there were very few working PAT cats. However, the activities of one of this rare breed were brought to my attention by his owners. Kittah, a seal point birman lives with Mary and Theresa Jones in Aberystwyth, Wales. He was the first PAT cat in Wales and from his own personal experiences probably knew all too well the need for comfort and support when you are not at your best. He was born with a hair lip and impaired vision. He also survived a severe bout of pneumonia which left him close to death on two occasions. To prevent a reoccurrence Mary knitted him a woollen all-in-one romper suit to keep him warm. His health worries, however, were not over. He had to travel to England to undergo surgery to correct the problem with his eyes. The operation proved to be a great success and now fully fit, Mary applied for him to become a PAT cat. He passed the health and temperament checks with flying colours and made his first visit to a local school for children with special needs. He was an instant hit with the children who stroked, cuddled and very happily played with him. He was keen to share his love and managed to spread his attentions equally amongst them.

The visit went very well but it wasn't until Mary went to the hairdresser of all places that she became aware of a significant after effect. A stylist at the salon that Mary had not met before described to her and two other elderly ladies the events surrounding a strange visit made to a school nearby. 'Do you know my neighbours' little boy is disabled and hardly talks but after being visited by a little old lady and her cat, he came home at the end of the day and wouldn't stop saying cat.' The stylist continued her tale and detailed how the boy's mother, who was very pleased, but a little puzzled by her son's behaviour and thought she may be imagining things, contacted his teacher who explained about the visit. As a consequence she got a kitten for her son and reports indicate that the new addition to the family was doing much to stimulate the boy's development.

Mary was obviously very pleased with what she heard and Kittah returned to the school for a further visit. Theresa, Mary's daughter, accompanied her with Kittah and observed the children's reactions. From what she witnessed on her visit, it seemed quite clear that Kittah's presence stimulated the children's senses. They mimicked some of his miaow noises and one wheelchair-bound young girl, who at first seemed unaware of the cat's activities, tried to raise her head when he was placed on her lap. Then, to the surprise of her carer, she rubbed her face in his fur. Theresa's experience suggested that the person who accompanies the cat also plays a very important role. She felt the children seemed more comfortable with her mother than they were with her. She thought this might be a result of them looking upon her as a teacher or parent figure rather than a kindly nanny or grandma.

Visiting cat programmes currently appear to be something of a rarity but the number of cats kept as pets is steadily increasing. More cats are kept as pets than dogs now, with the cat fast becoming the more popular companion animal for families where nobody is at home during the day. An Australian survey conducted by Cheryl Straede and G. Richard Gates MD in the early 1990s may offer some insight into the potential benefits of cat ownership. The study sought to investigate the relationship between psychological health and cat ownership. Ninety-two cat owners and 70 non-pet owning people completed mental health assessment questionnaires. The findings of the survey suggested that cat owners had significantly lower scores for general psychological health, indicating a lower level of psychiatric disturbance, and could therefore be considered to have better psychological health than the non-pet owning participants. I certainly find our present cat, Bebe, provides me with emotional support. If I'm feeling particularly wound up he inevitably makes his way on to my lap and, without thinking, I begin to stroke him. It certainly takes the edge off my day-to-day tensions.

Some would argue that cats' instinctive hunting of

mice and rats means they also make a practical contribution to human well-being. However, it seems extremely rare for a cat to offer practical support and assistance in a way similar to that derived from their canine counterparts. Nevertheless, the exploits of a now sadly deceased feline called Becky deserve the fullest recognition. Becky, a vivacious fluffy black and white rescue cat, lived with Angela Howard, a very charming work associate of mine, and her family for six and a half years. During this time she performed a dual support role which, to my knowledge, is quite unique.

Becky spent much of her time at Angela's home looking out for the family dog Boo Boo, a rather rotund and elderly shitzu. Boo Boo, who was both arthritic and almost blind, had great difficulty negotiating the garden steps which led on to a patio. Becky would wait for him just outside the door and then, as skilfully as she could, guide him along by walking before him and standing in front of any obstacle in his path. Once on the patio she used keep an eye on him while he mooched around and then she would guide him back in again on the return journey. Boo Boo often blundered into Becky on these trips but she never once became cross.

Becky was also very special to Angela's elderly mother, who was in her eighties. She, too, by very unfortunate coincidence, suffered from severe arthritis, was almost blind and extremely hard of hearing. Becky, however, helped her by providing some rather unusual sensory support. If someone called at the door she would tip her 'grandma' off by gently prodding her with a paw. Becky's actions, I am sure, enabled her to feel a little more at ease and in touch with her surroundings. Though for the most part a very lively cat, Becky was very calm and gentle with Angela's mother. This very sensitive cat always made a fuss of her when she stroked her and if she managed to feed her she would always give her a cuddle before eating the offering placed before her. Sadly, I never had the honour of meeting Becky but I have no doubt that she is very sorely missed.

In one particular case the fiercely independent nature of some cats provided quite unexpected therapeutic benefits. A mental health centre located in rural Michigan, USA, became the subject of a rather unusual and unplanned visiting therapy programme. A family of feral cats who had taken up residence nearby, began calling in on counselling sessions if the window in the therapist's office was left open. Generally the cats were welcomed into sessions, although if anyone was uncomfortable with their presence, they would be removed immediately. Before too long it became increasingly evident that the presence of these cats contributed to the therapeutic process.

Many of the clients who attended the centre quickly established strong, meaningful bonds with the cats, and by feeding and caring for them they did all they could to ensure their well-being. Not only did the presence of the cats enable clients to express themselves more freely, they also appeared to derive particular benefits as a result of the cats' wild nature. Eileen Wells and her colleagues, who reported these feral cat experiences, offer a number of explanations. All too frequently, the therapist's caseload included clients who, as children, had suffered sexual, physical and emotional abuse. Understandably, their perception of themselves as victims was often carried into adulthood and they experienced difficulty establishing and sustaining trusting relationships. However, these clients readily established attachments to the cats. They became particularly interested in the cats' ability to survive in bad weather and in the attentions of their natural enemies. It seemed as though the clients identified with the cats because they perceived them as also being victims. More importantly however, they also appeared to view them as survivors and, as a consequence, felt encouraged that they too could be survivors.

A number of the clients who had experienced profound difficulty with relationships throughout their lives became very attached to the cats. Some demonstrated for the first time ever the capacity for establishing a caring relationship. Eileen Wells and her colleagues believe this may have been because the clients perceived the feral cats, unlike domesticated animals, to be less contaminated by 'human identification'. They also feel that the feral cats enabled many of the clients to develop enhanced self-esteem. Unlike domestic cats, feral cats are not tame, trained, or even particularly attractive, but they demand respect for what they are, and live life very much on their own terms. For many people attending the centre, this has apparently proved to be a major initial step forward in changing their entire life view.

Recent research conducted by Mary Whyam and Liz Ormerod strongly suggests that a number of benefits can be derived from allowing companion animals in prisons. Their proposals for the extended implementation of animal programmes within penal institutions are supported by many in the prison service and by inmates themselves. Their survey of Scottish prisons identified cats, particularly feral ones, as playing positive roles in two special units for Category 'A' (top security) prisoners. In the Barlinnie Special Unit there are three cats Alfie, Lady and Ollie. Alfie and Ollie were born to feral prison cats. Alfie was hand-reared by his owner and a very strong bond of affection and trust has resulted. Lady, a rather attractive white cat, is owned by a man who has admitted that previously he hated cats. Pigeons are also kept in the unit. Prison staff believe the animals help normalise the institution and allow prisoners to openly express affection in an acceptable manner. The prisoners feel that watching the cats has taught them how to relax and that the animals are also good for the officers because they give them an interest too.

In another facility, the Shotts Special Unit, the resident cat was found as a two-week-old kitten after being abandoned by his feral mother. He was very carefully reared by one man and they too have developed a very close bond. The cat acts a little like a dog and follows his owner around. The cat is named 'A' Cat because his owner was an 'A' category prisoner who found him very soon after he had completed five years in solitary confinement. By all accounts, the adjustment can be terrifying and the prisoner concerned has stated that 'A' Cat saved his sanity.

Caressing and stroking are very much a part of any relationship with cats. Their soft fur, gentle nudges and propensity for filling our laps make them hard to resist. Touch and physical contact can be very important to our well-being and in this respect cats may play a particularly important role. An experience while working at a mental health unit in West London brought this point home to me more effectively than any academic exploration. One of the clients who attended the unit, Simon, had an anxiety disorder. I mentioned him in an earlier chapter because he was involved in the study I conducted with the dolphins. He was in his late thirties and, even with regular prompting, found it difficult to maintain what many would consider an appropriate level of personal hygiene so that his clothes were often grubby and creased. Simon was a great animal lover and particularly fond of cats. He often spoke about his own cat Jasper. He had also struck up a close friendship with a cat called Tonto who visited the mental health unit on a regular basis.

This very affectionate and incredibly fat tabby cat used the centre as a second home. He only lived across the road but I think he was attracted by our large garden and the enormous amount of attention he received from clients and staff alike. He liked the centre so much that it was often difficult to persuade him to leave at night when we were closing for the day. On more than one occasion he found a good hiding place and ended up being locked in for the night. I remember being woken from my bed very early one New Year's day by a phone call that required me to travel to the unit, with an extremely debilitating hangover I might add, to search and see if Tonto had been locked in. He hadn't been seen for over 24 hours and his owner had become very worried. Sure enough I found him skulking under a table looking a little bit sorry for himself. Nevertheless, he was a real character and a very welcome visitor. Simon would seek out Tonto whenever he could to stroke and talk to him. Tonto would return this affection by rubbing his head against him and would quite often lick Simon's hands and arms. Simon, who had few friends, clearly enjoyed Tonto's attentions and it was good to see him so relaxed and smiling.

The mental health unit attempted to offer an ever increasing range of practical support services for clients and one of these was subsidised hairdressing.

Christina, the hairdresser, visited the unit every Friday afternoon. She was also a fully trained therapeutic masseuse and would give a head and shoulders treatment on request. Simon, although almost completely bald, was one of her most regular appointments. He would have a wash and trim from time to time but he seemed to particularly benefit from the massage therapy. I assumed that he enjoyed the relaxing benefits of such a treatment, as many of us would, but for Simon it met another very important need of which I and the rest of the care staff were completely unaware. I only found out because Christina was exceptionally good at feeding back to me anything of significance imparted by a client while they were with her. Late one Friday afternoon, after she had finished all her appointments, she came to my office and said she wanted to share something with me that had occurred while treating Simon. He had commented to her how much he appreciated the massage therapy and went on to explain why he came to her for such regular treatments – sadly, it would seem that besides herself, Tonto and his cat Jasper, no one ever really made physical contact with him. She thanked him but didn't pursue the point. Maybe his isolated existence demonstrates how we humans, unlike animals, may be too quick to judge.

One of the most famous visiting animal assisted therapy experiences to date involved a young cat, but by anyone's standards she was no ordinary feline. CHATA, the Children in Hospital and Animal Therapy Association, organised an event where a five-month-old female tiger cub, Indi, was taken to visit children in Guy's Hospital in London. The visit in May 1997 was arranged as a means to offer chronically sick children on the wards a focus and something exciting to look forward to that offered some respite from their day-to-day hospital existence. After all, it's not that common to be treated to a private visit from such a rare and honoured guest. The event captured the public's imagination and it was a very hot item with the news agencies. Press and TV pictures of the sick children coming face to face with a tiger (although no direct contact was allowed) flashed around the world. The day's events were witnessed on the evening news by a seriously ill 17-year-old girl, lying in bed on a ward at University College Hospital, London only a few miles away from the scene of the wonderful encounter.

This young women, who I shall call Lizzie, was at a very low ebb after a painful and increasingly unsuccessful fight against leukaemia. However, she was quite taken with the idea of meeting a tiger so she asked one of her nurses, Jo, if there was any chance of the cub visiting her. The nurse was very happy to make some enquiries on her behalf because she felt this brave young girl certainly could do with a lift of some sort. Her long hospitalisation had taken its toll on her relationship with her family and the medical team treating her. Her eye contact, self-esteem and sense of control were all very low. Jo set about making some enquiries. Sandra Stone, the founder of CHATA, eventually received a fax from Jo in the form of an SOS explaining the situation. She was keen to help and cancelled a television appearance to enable her to make the necessary arrangements.

A preliminary visit to the hospital was made to talk through the planned visit with the nursing staff and microbiology specialists to ensure all necessary medical and health precautions would be taken on the day of the visit. Sandra was also keen to chat to Lizzie. She was located in a room which had to be accessed through two sets of secured doors due to the necessity for her to be in protective isolation. Sandra met up with Lizzie who was lying in bed. She had lost her hair and was extremely thin. It was very clear to Sandra, a nurse of many years, that she was seriously ill. However, their chat about the forthcoming visit appeared to cheer her up enormously. The staff on Lizzie's ward certainly noticed a difference in her during the period prior to the visit. She was keen to take back some control. She worked with the staff to sort out the times of her drug treatments so they didn't interfere with the visit. Her relationship with her family also improved. She made sure they were with her for the visit and arranged for her brother and sister to have time off from school.

On the day of the visit Sandra and her team had to carry Indi, the tiger cub, up three flights of the hospital's internal staircase to an enclosed area on the third landing. Lizzie was there in a bed chair accompanied by the necessary I/V drips and equipment. Her parents and brother and sister were also close by. The tiger was allowed to roam free and came very close to Lizzie, brushing against the side of her bed chair. She and her family were a little stunned by the animal's presence but they talked freely to the cub and to each other in a way that had become none to easy over recent months. There now appeared to be a closeness between them that had been missing for some time. The visit had been a great success and Sandra and her colleagues kept in contact with Lizzie, taking some of their more usual animals to see her on several occasions.

Several months later CHATA organised a trip to a wildlife park for over 250 patients and carers. The park was home to the now famous tiger and of course Lizzie and her family were invited along. Her parents decided, however, that Jo the nurse should be allowed to accompany Lizzie and her brother and sister, and she was very pleased to oblige. Lizzie was also very happy to take some control again and arranged pony rides for her siblings and chartered the route they would take around the animal enclosures. However, her condition by this time had further deteriorated and she was even thinner. Nevertheless she made every effort to participate in all the day's activities, smiling and talking quietly. Sandra recounted to me that Lizzie felt particularly honoured and privileged to stroke a new baby tiger cub, Inca, who was making his public debut. Another visitor to the park who witnessed their meeting asked Jo the nurse why she wasn't taking photographs of this special moment. Slightly embarrassed, she began to explain that she had taken so many pictures during the visit that she had run out of film. Lizzie's response to the concerned bystander left Jo too choked to speak, 'I don't need a photograph to remember this.'

Lizzie was discharged from hospital soon after the wildlife park visit. Shopping was soon replaced as her number one priority by frequent trips to the library and travel agencies to find out about tigers and their natural environments. However, Lizzie sadly died within a few months of returning home. The tigers had certainly offered some respite from her suffering and maybe, in just a small way, helped to bring her family together at the most painful and difficult of times.

The experiences of a dentist friend of mine also illustrate how the unexpected involvement of a cat can enhance the therapeutic process. Sharon used to run a busy dental practice in southwest London and was always keen to do all she could to reduce her patients' anxiety. Fully aware of this, her two practice nurses approached her with a proposition that she was initially very reluctant to consider. The proposition came in the shape of a rather dishevelled, elderly, long-haired black and white one-eyed cat. The seemingly stray cat had wandered into the reception and on inspection was found to have no collar and to be somewhat undernourished. He certainly seemed neglected and his longish hair was rather matted. He was in a very sorry state and welcomed the nurses' kind attentions. They decided to name him Harry.

The nurses very rapidly fell for the poor creature and were keen to take him into their care. They asked Sharon if they could keep Harry as the practice cat but at first she was not at all enthusiastic. The nurses, however, were very determined. They promised to take full responsibility for his welfare and after some begging and pleading they managed to persuade her that the potential benefits of him joining the team far outweighed any possible problems. Harry, they said, would be perfect for anxious children to cuddle and stroke before and during treatments, and he might also encourage them to attend for regular check-ups. This proved to be the deciding factor, and once Sharon agreed, they set about getting him back to full health. She made only one proviso, and that was that he would have to be cleaned up and thoroughly checked by the vet before he could join them.

After a few good meals, some veterinary attention and the nurses' love and care, Harry quickly returned to full health. He turned out to be a very loving and affectionate cat. He looked a little different, with just the one eye, but before too long he had made himself very at home in the practice and even Sharon had become quite taken with him. Well, the time had come for Harry to earn his keep and he certainly didn't let anybody down. At first his duties were restricted to the waiting room, where he proved to be very effective indeed in distracting both children and adults alike. His presence seemed to give the room a more relaxed feel and Sharon began to notice that her patients generally seemed a little more at ease.

Given the success of his work in the waiting room, Sharon felt that he was certainly up to taking on extra responsibilities in the treatment room itself. Harry was provided with his own special seat just a few feet from the patients' chair. He was encouraged to sit on his seat and if a child or adult appeared nervous, the attending nurse would ask them if they would like Harry on their lap to stroke while the dentist went to work. At first all went to plan and Harry became quite used to being placed on an anxious patient's lap. When he was not required he would sit perfectly well behaved on his special seat. Quite understandably from my point of view, his lap services were very popular so, more often than not, his services would be requested. Well I suppose a regular pattern had been established, so no one should have been too surprised when Harry began to make some clinical decisions himself.

A man in his early thirties came for an extraction and was shown into the treatment room. He gave Harry a gentle stroke but politely refused the offer of his company on his lap. The gentleman made himself comfortable in the chair and Sharon adjusted it to an almost horizontal position and then turned to pick up an instrument as the assisting nurse placed a bib around the patient's neck. The nurse withdrew but, as Sharon turned to begin the procedure, I guess Harry decided the prostrate patient did need his support after all. Sharon and the nurse could only watch in horror as he leapt from his chair on to the unsuspecting patient's lap. The man, completely unprotected, took a blow that shocks the system in a way that only men know and then matched Harry with an impressive leap of his own. The cat, now in a state of shock himself, rapidly scuttled from the room through the half open door into the waiting room.

Luckily Harry had avoided a direct hit and hadn't used his claws as brakes. The only slightly sore but very shaken patient turned to Sharon and her nurse. It was as though they had been struck dumb as they faced him speechless. His anguished expression however, slowly dissolved into a smile, which in turn gave way to laughter. He was soon accompanied in mirth by his attending physicians who, now finding their speech restored, couldn't apologise enough. Once calm had returned, Sharon efficiently completed the extraction. As this most understanding of patients was about to leave he thanked Sharon and the nurse and offered these words on parting, 'I must admit to having been a little tense but that's one heck of a way of breaking the ice.'

Harry repeated this trick on only one further occasion and thankfully, again, no permanent damage was done. For the most part, however, he performed his duties admirably and spent a number of years keeping the practice calm.

CHAPTER 8

THE UNSUNG HEROES

An enorm range of animals have demonstrated qualities that have offered therapeutic support and comfort to their human companions. So far within these pages I have concentrated on some of the most familiar. However, I would now like to turn my attentions to the less publicised contributions of some of the minor players.

A very recent addition to the therapeutic family is the ferret. June McNicholas at Warwick University is now using them in her work with children with autism. This condition is a developmental disorder which isolates the child or adult from the world as we see it. Autism impairs the natural instinct to relate to fellow human beings and words, gestures and facial expressions can mean little to someone with this condition. They show scant curiosity or imagination and frequently seem indifferent to the usual process of two-way communications. June had found in her previous research that, just occasionally, a child with autism can relate to an animal in ways that they are unable to with people. She began working with ferrets in the hope that they may act as a catalyst for communicating with some autistic children.

Her initial suggestion to add a ferret to her university's animal therapy team, which included trained dogs and docile guinea pigs and rabbits, sent her colleagues into a state of shock. However, she firmly believed they would be a useful addition given that they are friendly, curious, interactive, comical and furry. Eventually she persuaded the other members of her research group to at least meet her two ferrets, Bracken and Wombat, before they closed their minds to the idea. June describes Bracken as 'a huge sandy teddy bear who chuckles and dances continuously', whilst Wombat is 'a calm, gentle, silver mitt, whose main thought in life is to find a lap to sleep on'. The charm offensive worked and following the meeting they were very readily welcomed into the team.

Wombat was selected for some tests with an autistic child. The young girl had shown little interest in the researcher's dog and although she had interacted with their guinea pig this had revolved around feeding it carrots which has its obvious limitations. However, as soon as Wombat was taken out of his box, she was instantly taken with him. She stroked him, spoke to him and placed him gently on the floor where she crawled alongside him talking all the time. She also took him to her mother to 'share', which is very unusual behaviour for an autistic child. The young girl was closely supervised by June, but she was very gentle with Wombat throughout their encounter. When it was suggested that he should be returned to his box, she made it very clear that she didn't want him to go by running to the box, shutting the lid and sitting on it.

To bring the session to a close Wombat had to help the girl on with her coat and walk with her to the car. She kissed him goodbye and waved. During her time with the ferret the young girl displayed a number of behaviours not typical of those she showed with people. The research in this field is very much in its early stages and at present it is not known why animals can sometimes have this effect. Wombat had clearly made some impact on the child and future studies are planned.

June has used two of her other ferrets, Jack and Robyn, to work as therapists in a rather different situation. She has taken them on to children's hospital wards in an attempt to reduce the children's anxiety about forthcoming surgery. In addition to being a lot of fun and a distraction, these two very lively customers had at the time some visible attributes that helped to make an important point. Both Jack and Robyn were recovering from operations themselves and had very evident scars and stitch marks. Their boundless energy clearly demonstrated to the children that they, too, would soon make speedy recoveries from their surgery and be up on their feet in no time.

As far as I am aware, June's work is the only animal assisted therapy programme that utilises ferrets. She hopes to continue her work into the future but for now, since ferrets have a reputation for viciousness, she feels it makes a nice change for them to be seen as the gentle, trustworthy animals they can be.

I now turn to an animal that holds a special place in many people's affections, including my own – the donkey. As long as I can remember, just coming face to face with one has always brought a smile to my lips. For many, these endearing creatures also hold religious or spiritual significance, which further enhances their appeal. The Elisabeth Svendsen Trust for Children and Donkeys has utilised the animal's small size and placid nature to develop an alternative to the more traditional horse riding therapy. The Trust was established in 1989 and now has two centres in Sidmouth, Devon, and Sutton Park, Birmingham. Their main aim is to offer children with a range of special needs and disabilities the opportunity to enjoy the pleasure and satisfaction of learning to ride in a supportive and caring environment. Children with learning difficulties and physical disabilities attend the centres, as do children with visual or hearing impairments. Wherever possible, the children's activities are tailored to their individual needs. In the summer the Trust operates mobile units which take donkeys to children in schools and hospitals who, because of their disabilities, are unable to travel to the centres.

Up to 150 children attend the centres each week. They range in age from one year olds to young adults. Any of those attending who weigh over 50 kg (8 stone/112 lb) are too heavy to ride donkeys, so they are taught to drive a small cart. Riding and driving skills are taught by qualified instructors. The children are also encouraged to learn how to groom and care for the donkeys. The staff have found that the setting of readily achievable goals in each riding session increases the motivation of the children and raises their confidence and self-esteem. Riding can also improve the balance and dexterity of many children. Donkey riding offers similar therapeutic benefits to those derived from horse riding. Improvements in learning and speech are common. Physical gains are also derived from this version of mobile physiotherapy.

One little boy's mother has commented to the Sidmouth Centre staff that on the days he rode the donkey he did all his physiotherapy without even noticing. She added that he thoroughly enjoys his riding and, unlike conventional physiotherapy, it has never felt like a chore. Ben, a young boy who suffers from very stiff limbs, is particularly motivated by the donkeys. His mother has reported that on the days he knows he's going to the centre, his limbs appear far more supple and his dressing is done in half the time. The donkeys have also helped to calm some of those with more challenging behaviour problems and are often very responsive to an individual child's particular needs. One of the exercises undertaken with the children involves posting a letter. Each child reaches from the donkey to a pile of letters, picks one up, holds on to it until they reach the post box, and then leans over and posts it. One very small boy just could not grasp the letter from the pile however hard he tried. Pedro, one of the more experienced donkeys who was well used to the letter posting routine, reached around and picked up the letter for him. He held it in his teeth until they reached the letter box then stopped and gave the letter to his leader who passed it to the little boy who was then very delighted to post it.

Donkeys offer a smaller scale but equally effective riding therapy as their equine cousins. They may be just the thing for children of all ages who are a little frightened by the size of horses. My old fear of horses didn't extend to donkeys and they continue to be a personal favourite.

One of the more unusual and, to some, more controversial forms of animal assisted therapy involves the training of small capuchin monkeys to support quadriplegic people in their daily living. A French programme – Programme *d'Aide Simienne aux Personnes Tetraplegiques (PAST)* – was established to place these monkeys with a number of people as a possible means to reduce required levels of human assistance or to replace the use of special robots. The instigators of the programme believed that placing a capuchin monkey with a quadriplegic person could increase their independence, and lessen feelings of loneliness through the development of a caring and affectionate relationship.

The programme team were keen to ensure the welfare of both the monkeys and the quadriplegic people involved in the project. To this end, an ethical committee was established comprising the project team members, representatives from the local branch of the French Society for the Protection of Animals, families who would foster the monkeys in the early stages of their training and local zoological parks. Quadriplegics, psychologists and a veterinarian also joined the committee.

Individuals selected for the programme were functioning quadriplegics, who used electric wheelchairs and had endured more than two years of disability following diagnosis but who were otherwise in good health. They were unemployed but living independently, even if this necessitated the services of a care assistant. They had to be able to speak as loudly and clearly as possible. It was also extremely important that the would-be monkey recipient was very well organised in their adjusted life style.

Capuchins were chosen as the most suitable monkey for the programme because their skills are well documented. The species is small, highly manipulative, highly social and has well developed learning capabilities that are closer to those of the great apes than other monkeys. The capuchin's induction programme had three main phases. The first – the socialisation – started when the monkey was weaned at the age of two or three months and placed in a foster family where the human mother performed a substitute or surrogate mother role. This phase lasted until the capuchin was four years old.

The training phase lasted from six to eight months and took place in a rehabilitation centre. The training methods involved the use of positive food rewards and responses to encourage appropriate and desired behaviour. Negative responses were limited to verbal commands only. During this phase the monkey played with capuchin partners, roamed freely in a room with a trainer and received two 40-minute training sessions.

The final phase was the placement itself. The quadriplegic person was instructed in all aspects of his relationship with his new companion, including how to give orders and gain respect. They were also taught about capuchin behaviour and how to instruct other people on how to behave with the monkey. Care assistants were also introduced to the capuchin and they received special instruction on how to care for the monkey and protect the capuchin's relationship with its human companion. Recipients were encouraged to view their monkey companions like another person and respect their need for their own objects and personal space.

One particularly controversial aspect of this programme was the decision to file all of the monkey's canine teeth so they were the same size as the incisors. Additionally, if a biting problem occurred, recipients could approach the ethics committee to have their capuchin's teeth extracted. Such a decision, however, would only be made following the most thorough of investigations.

The programme successfully placed seven capuchins and the recipients' own reports indicated that the project had been a great success for them. Nearly all recounted that their lives had changed for the better and in some cases far more than they had ever hoped. Only one person reported problems controlling their capuchin, but this appeared to be the result of a lack of commitment on the part of the recipient. However, all recipients informed the project team that their level of activity had increased and that a caring relationship had been established. As a result, they all felt less lonely than previously. Similarly, recipients reported feeling more independent and more comfortable when on their own. Some also reported that people responded to them more positively when they had the capuchin with them.

The French programme was based on similar projects that had been established in the USA and Israel for sometime. However, the French programme had the joint purpose of producing an objective assessment of the project. The French team are hoping to produce this full evaluation in the not too distant future and I will await their findings with some interest. I do appreciate, however, that some may feel that the placing of non-domesticated animals in human living environments in this way is open to question.

An American study in the early 1980s concluded that watching fish in an aquarium decreased blood pressure to below the level of a person who is sitting comfortably in a chair and resting. Watching the fish was also found to produce a state of calm relaxation. Another study conducted by Carol Riddick in the mid 1980s in Maryland, USA, suggested that having an aquarium in the home may benefit older people. A group of elderly residents of a housing project were each given a fish aquarium. After six months their broad levels of physical and emotional well-being were compared to groups of non-aquarium owning elderly people. Comparisons were made on blood pressure, levels of loneliness, happiness, anxiety and leisure satisfaction. The research findings indicated that those with an aquarium experienced a decrease in one measure of blood pressure (diastolic) and an increase in leisure satisfaction related to relaxation, whereas no changes were noted for the non-aquarium owning group whose social situation had remained unchanged.

These studies are but two examples of a number of

research projects which have indicated that the company of fish can benefit us in terms of reduced blood pressure and improvements in our psychological health. However, I was somewhat surprised when I came upon some fish that offer a rather different approach to enhancing our well-being. I am speaking of the healing fish of Kengal. These fish live in mineral-rich pools fed by a hot spring located in Ilce near Kengal, about 400km (250 miles) from Ankara in Turkey. The hot spring fills several pools with warm water at 36°C (97°F) that contains literally thousands of these tiny fish, ranging in size from 1–2 cm up to about 9 cm (½ – ¾ inches to 3½ inches). People from all over Turkey and Europe visit the springs to treat a variety of skin conditions, ulcers and rheumatism. However, the majority of the foreign visitors come seeking a cure for psoriasis. Three species of small fish live in the pools and each in turn performs an active role in the treatment of this most stubborn of conditions. Firstly, the species of fish known locally as strikers, continuously strike at the scales of skin on the surface of the affected areas. Their actions to soften the diseased skin are helped by the hot mineral-rich water. This part of the treatment can last up to five days or more ensuring that the skin is soft enough to allow the second wave of fish, the jabbers, to do their work. Their part in this process is to puncture the affected skin and draw blood. Little blood is actually spilt because the third species, the lickers, set to work almost immediately. Their tongues secrete a healing and clotting agent which stops any bleeding.

In most cases, the treatment lasts up to 21 days and requires that one spends up to seven hours a day in the pools surrounded by the fish – not a treatment that would suit everyone. Anecdotal reports and stories claim the fish can cure psoriasis completely, but I was unable to find any research either supporting or challenging these assertions. As a psoriasis sufferer myself although, thankfully, only mildly affected, I'm extremely interested in possible new treatments. My experiences bathing in the Dead Sea have certainly convinced me of the beneficial effect of immersing oneself in mineral-rich waters in a warm and sunny climate. Whether a cure can be affected with the help of these fascinating fish, well, one can only hope.

The doctor fish of Kengal are also reported to offer another service. If you are prepared to open your mouth on the surface of the water, the fish will swim in and clean your teeth and gums. Even with my levels of anxiety about dentistry, I think I'd rather take my chances with Harry in the dentist's chair.

The spa at Kengal also claims a cure for the potentially fatal condition of erysipelas, otherwise known as St Anthony's fire, an infectious disease of the skin. Sufferers enter a nearby stream which is also fed by the hot springs and is home to healing eels. The stream's inhabitants measure between 25 and 35 cm (10 to 14 in) long. The eels are said to pick up the scent of someone afflicted with the condition and then swim to their aid. They circle their patient several times then swim away. They then return and pierce the skin to draw off the inflammation. The treatment is then said to be complete and the patient cured.

I would strongly recommend that anyone contemplating a trip to Kengal for treatment for either condition should firstly consult their doctor. However, I fully intend to give it a try at some point in the very near future.

Rightly or wrongly, I have always associated budgies in

cages with elderly people. They are, however, relatively easy to keep and can be very good company. In the mid 1970s a British research team led by Roger Mugford conducted a study to evaluate the possible benefits to elderly people of keeping birds at home. Thirty elderly people aged between 75 and 81 were divided into several groups. Two of the groups were given parakeets, another two were given flowers, begonias to be precise, and one group was given nothing. At the beginning and end of the study all group members were asked to complete questionnaires focusing on their thoughts and feelings about themselves and other people, and their living environments. The study found that the people given the birds showed markedly positive attitude changes towards themselves and others. They had also formed strong emotional bonds with the birds. The parakeets certainly seemed to have had a very positive effect on their lives and these people reported improved psychological health. The birds also became new and interesting topics of conversation for them and as a result helped them to socialise with neighbours more.

In the not too distant past I was asked to re-home Noel, a cockatiel whose elderly owners could no longer look after him. I asked the residents of a mental health housing project if any one of them would like to care for him. A group of them agreed to share the responsibility of looking after the bird and he was placed in the communal dining room of their group home. This nine-bedroomed house offered supported community living for people with mental health problems who, for a range of reasons, were unable to live alone. For a number of the house residents difficulties relating to isolation and loneliness had been a major consideration in their decision to move into a shared environment.

The bird seemed to settle into his new home quite well and certainly added a new dimension to the living environment. However, after about four or five weeks, I was approached by the residents because they had become a little concerned about the bird's well-being. Even though they very much enjoyed having Noel in the house, they were worried that he would become lonely on his own without other birds for company. To be honest, I hadn't considered Noel's needs in quite this way and I was very pleased that they had brought their concerns to my attention. After some further discussion, the group decided that he should be taken to a local aviary where he would never be lonely. Within a week or so he was moved to his new home and all reports indicated that he had settled in well.

I certainly believe that the residents' selfless empathy with the bird's plight was born out of their own painful experience and should be commended. In a similar way to the feral cats I discussed in the previous chapter, the cockatiel seemed to offer an alternative means for reflecting on difficult experiences, which can prove an invaluable first step on the road to recovery.

I would like to devote the remainder of this chapter to the work of a rabbit and a guinea pig in the employ of the Children in Hospital and Animal Therapy Association – CHATA. These small animals have brought an enormous amount of support and comfort to children over recent years and deserve some special recognition. The same could be said for Sandra and Ronnie Stone who run the association. When I met with Sandra to record the exploits of Obi, the rabbit, and Dolly, the guinea pig, some of the realities of working with profoundly sick and terminally ill children began to hit home.

Lucy was admitted to Great Ormond Street hospital at the age of six weeks following a road accident. At first she was not expected to survive and after her religion had been established, she was very quickly baptised. However, she rallied and managed to hold on to life. Nevertheless, she remained seriously ill and a long period of hospitalisation was expected. Her mother, a single parent who was also seriously injured in the crash, was to suffer an indefinite stay in hospital. (Sadly, she never recovered enough to take care of Lucy herself.) It therefore appeared that, apart from the Great Ormond Street staff, there was no one to visit or attend to this very young infant as her maternal grandparents felt unable to offer her the support she would surely need as time went on.

The hospital decided to advertise for potential foster parents and soon a couple who were prepared to take on the role came forward. They visited Lucy on a regular basis and bought her presents on her birthday and at Christmas. Her condition had stabilised to some degree, but as a result of her injuries, she was paralysed from the waist down and had to be fed intravenously. Unfortunately, and for reasons unknown, her foster parents relinquished their role when Lucy was only two years old. Thankfully, a new couple were found and they became absolutely devoted to her. Emotionally, things were looking up a little for Lucy but she was now going on for three years old and she had never left the hospital. It was at about this time that Sandra met her. Lucy loved the CHATA animals, who included at this time a mongrel dog called Sandy and a variety of rabbits and guinea pigs. Sandra described Lucy to me as an adorable, plucky, outspoken little girl with the most wonderful cockney accent which she had picked up from the nursing staff and other patients.

Lucy found it difficult to contain herself when a CHATA visit was imminent. The staff would position her by the window so she could see them coming. They kept a very close eye on her, however, because even though she was semi-paralysed, she became so excited, that she would hang almost half way out of the window waving to them on their arrival. She would yell at Sandra across the courtyard to find out which animals she had brought and to remind her that it was her job to open the carriers. She wanted to take some control, which is more than understandable given her level of dependency.

This very brave child underwent numerous operations and even immediately following these procedures she was keen to see the animals. The staff did all they could to accommodate her wishes and they set aside a little ante room where Sandra could see her and she could be with the animals very soon after surgery. Lucy always insisted on cuddling them whatever condition she was in. It didn't matter what procedure she had just undergone or whether she had drips attached, she wanted to hold the animals.

Sandra and the animals visited Lucy on a regular basis for four years. During this time she left the hospital only once, to spend Christmas with her foster family. She underwent a number of further surgical procedures, including an unsuccessful liver transplant. This most courageous child endured an enormous amount of pain and discomfort. Her condition continued to deteriorate and a bowel transplant was considered in a final attempt to save her. However, in the end, the doctors felt she had suffered more than enough and, given that there was very little chance of success, the idea was abandoned. As she became weaker she was unable to cope with Sandra's lively dog but she became particularly fond of an albino rabbit called Obi. He had a wonderfully docile temperament and he would sit for hours with children just like a soft toy. Lucy wanted him with her all the time and they developed a very close bond. Obi gave Lucy tremendous comfort towards the end of her life and he also became very special to her foster parents.

Lucy died when she was seven years old. Her foster family asked Sandra if Obi could attend both her funeral and a memorial service. His presence did much to comfort those attending these services. The chaplain who had baptised Lucy as a very young baby conducted the memorial service, and he was very deeply moved. He commented to Sandra that he had drawn a good deal of comfort for himself by just looking at the congregation and seeing Obi and knowing how much he had meant to Lucy. It seemed that Obi gave everyone attending that day a cuddle as they remembered a very brave child.

The foster family approached Sandra for a photograph of Obi and she readily obliged. They made the request to enable a stonemason to make a gravestone in his image. Lucy is buried in a little garden area of a cemetery in Kent with an Obi the rabbit memorial stone at her head.

Sandra was introduced to a teenage girl Helen who had been admitted to Chase Farm Hospital in London suffering from anorexia. She was very angry with the world and, as is often the way with this condition, she would sometimes indulge in self mutilation. When Sandra first spoke to her she asked if CHATA brought cats to visit because she had a cat at home. She was a little disappointed when Sandra explained that this wasn't possible but she became quite interested when a

guinea pig was suggested as an alternative. Before too
long she was introduced to Dolly. Dolly had a long coat
and behaved in a very responsive way making little
chatting noises. Helen found her very appealing and
took a liking to her on their very first meeting. Within
a few weeks she began to bond with the guinea pig and
very much looked forward to her visits.

Anorexia is a complex condition and the staff of the
hospital did all they could to offer support. They
adopted a sympathetic approach to her situation but
they seemed to be making very little headway. They
then decided to try a different method of working.
The staff were very aware of how much she looked forward to the CHATA visits, so they decided to use
animal therapy as a means of motivating her. In simple
terms they explained to Helen that if her condition
improved and she ate a little, the animal visits could be
extended. However, if her situation remained
unchanged or deteriorated, the visits might have to be
shortened. This method of working is not always successful but Helen tried very hard and it worked
extremely well for her.

Thankfully, her condition continued to improve and
she went on seeing Dolly on a regular basis. They had
built up quite a relationship when Sandra had to break
the news to Helen that Dolly herself had become very
sick and died. Not unexpectedly, she took it quite
badly. The ward staff were a little concerned but Helen
managed to cope with the loss and was keen to see the
other animals. Her bond with them may not have been
as strong but Sandra had built up a close relationship
with her and continued to visit on a regular basis. It
was during this period of time that CHATA was being
registered as a charity. Sandra had become aware that
Helen was a very gifted artist, so she asked her if she

would like to be involved in designing the Association's logo. She rose to the challenge with a good deal of enthusiasm and her drawing was adopted as the logo.

CHATA clearly played an instrumental role in Helen's recovery from anorexia. Sandra still keeps in touch with her and by all accounts she is doing very well. She is studying at Oxford University, eating very healthily and looking beautiful.

As you will discover in the following pages, guinea pigs and rabbits can also play a pacifying role in crises.

CHAPTER 9

THE 'A' TEAM

A range of animals have clearly demonstrated their ability to offer therapeutic support and comfort in the most challenging of environments and situations. Their work has encompassed both crisis interventions and the establishment of ongoing relationships. Many of these endeavours are characterised by the ability of animals to affect change or offer hope where human efforts have been unsuccessful. To illustrate how animals can intervene in what could be described as emergencies, I shall now offer an account of the activities of Alf the guinea pig and Zowie the rabbit.

The first point I should make is that officially they didn't exist. They lived in a home for young adults with a range of conditions, including autism, who exhibited challenging behaviour and who could sometimes be aggressive. The home, in north London, was managed by Julia McAvoy, a colleague from my mental health work whom we met earlier in the opening chapter. She and her staff had noticed that many of their residents were at their most relaxed when looking over the fence into the next garden and watching a neighbour's rabbit playing in a wooden hutch and run. Julia made what she described to me as an executive decision – that some pets would be bought and kept in their garden. The keeping of pets in the house, however, clearly breached departmental standing orders but since, technically, the garden wasn't the house, she felt happy to proceed. So the staff set to work building a hutch and a very extensive run to accommodate some smallish animals and to prevent them getting into neighbouring gardens.

Julia went to an RSPCA centre to see what animals

they had that were in need of a home. A guinea pig caught her eye and she thought he would be quite easy to take care of, but she was informed that he couldn't be separated from his best friend, a rabbit, so she took them both.

Many of the residents of the home had great difficulty coping with change and at first they seemed a little unsettled by the arrival of Alf and Zowie. However, they soon adjusted to their new companions and became very at ease with their presence. Before too long, the residents would begin their days after breakfast by sitting in the conservatory of the house watching the rabbit and guinea pig in the garden. One of the members of staff then began opening the conservatory door in the morning and placing a tray of carrots in the doorway to encourage Alf and Zowie to come a little closer. After about four days they started coming into the conservatory and then, quite unexpectedly, they began climbing and jumping into people's laps. At first Julia was concerned that some of the residents might be frightened and mishandle them. Her worries, however, were unfounded. They began to gently touch, caress and stroke the animals and this is how things were to continue. Strong bonds were formed as the residents learnt to care for them and these small creatures seemed very happy to be in their charge. The rabbit and guinea pig visited the house every day.

One resident of the house, Stephan, a rather large young man who stood over 1.8 metres (6 feet) tall, was often extremely difficult for staff to work with. When he was feeling unhappy he could become very aggressive and violent. He would grab members of staff, sometimes ripping their clothes, and then sink his teeth into their flesh. Every single staff member had required hospital treatment for bite wounds following one of his outbursts. They continued to work with him because they felt he had been maltreated in the past and that, with time, they might be able to help him express his emotions without causing harm to others.

Julia and one of her colleagues had noticed that Alf had a very soothing effect on Stephan. If he was on his lap, he would laugh and pet the guinea pig. At first they were unsure whether at some point he would change his mind about Alf but it soon became apparent that he adored him and not once did he ever become aggressive with him. However, his aggressiveness towards staff continued to be a problem. On one never to be forgotten day, Stephan was particularly low and not unexpectedly attacked one of Julia's team. He had pinned the person down and was sitting on them when Julia entered the room. She knew it was only a matter of time before he would bite, and given the position he was in, the throat or face seemed the most likely target. Julia very quickly wrapped a towel around her arm thinking that if she could get her arm between him and person's neck, he would bite that instead and she might be able to release the trapped staff member. Then, just as she was about to intervene, it suddenly occurred to her to pick up Alf. Luckily he was in the conservatory, so she put him under her arm and then ran back to the scene of the attack. She put her arm in front of Stephan's mouth, he immediately sunk his teeth into the very thick towel and, amazingly, bit right through. He was now biting into the bone of Julia's thumb. In absolute agony and with tears rolling down her cheeks, she somehow managed to place the guinea pig close to his right hand. She then repeated over and over, 'There's your guinea pig, there's Alf, look at Alf, look at Alf.' Eventually, without turning his head, he took his right hand off the pinned down staff member and began to stroke Alf. Gradually he moved his head so he could see the guinea pig. He then let go of Julia's thumb, released her colleague and sat on the floor stroking the guinea pig. More quickly than ever before the outburst was over. Thanks to Alf, Julia kept her thumb and thanks to Julia, her staff member escaped a potentially life threatening injury.

The guinea pig proved to be consistently effective in calming Stephan and, as a result, he slowly became more able to deal with his anger without inflicting injuries on others. He has now moved on to live in a house where he has almost sole responsibility for the care of several guinea pigs.

Zowie the rabbit also had the ability to distract and calm. Another young man, Ralph, when anxious, worried or angry, would shout and scream while breaking windows, tables, chairs, televisions, in fact anything that was not fixed down. The staff were often at a loss as to what to do with him. In the main they would retreat from the room and just wait for him to calm down. However, by pure chance, the staff team realised that Zowie could provide a means for dealing with these extremely destructive episodes. On one such occasion, Ralph was in full swing and Julia felt that they must try to do something because broken items could not be continuously replaced and, more importantly, Ralph might hurt himself. So, with one of her colleagues, she entered the room he was in the process of destroying and somehow managed to coax him into the conservatory. As they entered, Zowie began hopping around in the garden at an ever increasing rate. Ralph's attention was slowly distracted as he watched the animal through an open door. The rabbit continued to hop faster and faster and within minutes Ralph was beginning to relax. The rabbit, who now had his full attention, appeared to be mirroring Ralph's level of activity and as he calmed, she began to slow down. When he became completely still, Zowie slowed right down and then walked over to him, stopped at his feet seemingly waiting to be picked up. This was slightly unusual in itself because, even though she was a very friendly rabbit, she rarely made the first move or instigated encounters. Ralph picked her up, stroked her for a moment or two and then put her down. He then very quietly went back inside the house. Zowie had brought one of Ralph's angry episodes to a conclusion in a matter of minutes, a process that in the past had taken up to two or three hours.

Julia, who to this day feels that the rabbit had acted intentionally to soothe Ralph, witnessed Zowie respond to him in the same way on numerous occasions. The rabbit's reaction was so similar each time that she was able to predict when Zowie was going to speed up, slow down and which route she would take around the garden. Julia was not alone in witnessing this. Several other staff members would watch with her and they would all look on with some amazement as the rabbit repeated her established pattern. The finale would always be the same too. Zowie would come over to Ralph and sit at his feet for about thirty to forty seconds before he would bend down, pick her up and stroke her once on each ear. He would then put her back on the ground and go inside and sit down quietly.

In terms of man-made environments, some of the most challenging exist within the penal system. In an earlier chapter I briefly referred to the study of animal programmes in prisons conducted by Mary Whyam, a senior probation officer, and Liz Ormerod, a veterinarian. Their work has left me in no doubt that in these most uncompromising of surroundings, animals appear to provide a range of tangible benefits for inmates, prison officers and society at large.

Probably the most famous example of animals having a positive effect in a prison environment concerned the story of Robert Stroud, more commonly known as the 'Birdman of Alcatraz'. After committing two murders and narrowly avoiding execution, he spent over 20 years, from 1920 to 1942, incarcerated in Leavenworth Prison. He had been without doubt a violent man but during this period he began to keep birds and some reports indicate that as a result, he experienced a profound change of character. It was not unusual at this time for prisoners to keep birds but at one point he kept nearly 300. He was given two extra cells to accommodate them. Stroud used his time to conduct detailed research on his charges and produced two books, *Stroud's Digest on the Diseases of Birds* and *Diseases of Canaries*.

Much of the pioneering work in this field has been conducted in the USA. The first structured animal programme in a maximum security institution was established in 1975 at Oakwood Forensic Center in Lima, Ohio. The programme was initiated by David Lee, a psychiatric social worker attached to the centre, after he had noticed improvements in a small group of inmates who had cared for an injured bird. As a consequence, he conducted a year-long comparative study between similar groups of prisoners with and without pets, on two different wards. The findings of the study were quite startling. The patients on the ward with pets required half the normal amount of medication. There was also a marked reduction in violence and no suicide attempts were made during the year. The non-pet ward had eight documented suicide attempts. These convincing findings were recognised by the prison authorities and as a result, similar animal programmes have continued at the centre.

A dog care programme was established at Purdy Women's prison in Washington state in 1981 by Kathy Quinn, a former inmate. The programme centred around the rescue and rehabilitation of stray dogs by the prisoners. The scheme has been enormously successful with over 500 dogs' lives being saved. A number of the dogs have been trained to work as assistance dogs for people with severe disabilities. The Purdy programme has been replicated in other parts of the USA, Canada and Australia. The following are few examples of subsequent programmes.

Dr Earl Strimple developed the PAL programme at Lorton prison, Washington, DC, in 1982. The programme placed a variety of pets – cats, birds, fish and rabbits – with prisoners and a club was established to offer animal care instruction. A training course in animal health technology was also developed. The benefits of PAL soon became apparent. The re-offending rate of men who were involved in the programme was 13 per cent compared to the US national average of 62.5 per cent.

A female prisoner involved in the Prison Pet Partnership Program in Gig Harbor, Washington, who suffers from epilepsy, is performing a most valuable role at the cutting edge of animal assisted therapy. She is working with service dogs and if one of them alerts her to a seizure, they are placed with other epilepsy sufferers. Up until May 1998, the placed dogs have had a 100 per cent success rate at predicting their new owners' epileptic seizures.

Mary Whyam's and Liz Ormerod's detailed examination of animal programmes in American prisons strongly suggests that both humans and animals generally benefit in as much that: *i)* animals help and enhance relationships between prisoners and between prisoners and prison staff; *ii)* prisoners' self-esteem is enhanced by them becoming animal caretakers; *iii)* work with animals can be translated into a marketable skill on release; *iv)* animals receive a high level of care as prisoners develop the human traits of empathy, compassion, patience, responsibility, trust and nurturing; and *v)* the animals are trained and domesticated so that they can be easily adopted.

Specific human benefits include a reduction in violence, less medication for prisoners, fewer suicides, reduction in illicit drug taking, an improved perception of the prison by the community and, most importantly in terms of society at large, a reduction in the rate of re-offending.

Their own research in Scottish prisons, which involved both inmates and prison staff, identified a number of related benefits that included an increase in the level of communications between prisoners and again between inmates and staff. Visitors seemed more relaxed and stayed longer. They also found that the presence of animals resulted in a reduction of staff stress levels.

Mary Whyam conducted a survey of English and Welsh prisons in 1995 and found a number of established animal projects. The most popular programmes involved birds and fish. Budgerigars, canaries and cockatiels were the pets most frequently kept. Several prisons had a duck and one institution boasted a visiting pair of falcons. A number of jails kept a prison cat. Several establishments were a home to farm animals, horses and donkeys. One prison had a visiting PAT Dog scheme.

She also identified a number of community based projects. Some prisoners were allowed to visit local schools with farm animals and one prison organised work placements on a community farm. Several jails supported Riding for the Disabled Groups. These offered prisoners the opportunity to perform a useful community service which also did much for their self-esteem. A number of prisoners were also employed in animal sanctuary work.

Differing groups and individuals hold a range of views and attitudes on the constitution and objectives of the penal system. Many would advocate containment and punishment as the primary considerations while others would place greater emphasis on the rehabilitative process. Personally, while acknowledging the importance of all these elements, I believe the effectiveness of any system may ultimately be judged by the levels and nature of re-offending following release from custodial sentences. Research certainly suggests that animal programmes positively impact on re-offending rates in addition to the other more immediate benefits.

My exploration of the benefits of having animals in prisons would be incomplete without me outlining the views of prison staff and inmates on what underlies these positive effects. Mary Whyam and Liz Ormerod sent questionnaires to the staff of 21 Scottish prisons and 25 prisoners participated in discussions on the subject. Responses to the staff questionnaires identified four major factors: *i)* time – the keeping of animals is a structured activity which provides interest and constructively occupies time; *ii)* therapeutic value – the programmes seem to relax inmates. The presence of the animals also creates a more convivial and relaxed atmosphere for staff and prisoners; *iii)* normalisation – the programmes enhance the prison environment by providing an element of normality; *iv)* nurturing – the animals provide prisoners with something to care for. Inmates develop more caring attitudes and the animals enable prisoners to express affection in what is considered an acceptable manner. The prisoner discussions revealed similar factors. Although less emphasis was put on time structuring and normalisation there appeared to be a consensus on the importance of nurturing. For the prisoners the most important factor was the therapeutic value.

Mary Whyams' and Liz Ormerod's exceptional research in this field is most persuasive. I just hope for all of our sakes that the powers that be are taking note.

After ten years of work in a number of community based mental health projects, I am no stranger to the large scale 'de-institutionalisation' of care process of the 80s and 90s, more commonly known as Community Care or Care in the Community. It seemed that many of those involved in these proceedings spent more time identifying the most appropriate language rather than focusing on the substance of one of the most significant changes this century in our approach to the support of some of the most vulnerable members of our society. It would have also been nice if the government of the day had sufficiently established community resources before they began closing the larger institutions, but I guess you can't have everything. I mention all this as a means of introduction to the story of a young golden retriever who probably did more to enable care in the community than a number

of designated national and local government representatives I could mention.

In the early 1980s, a so-called purpose designed
hostel for mentally handicapped people was built in
the middle of one of northwest London's roughest
housing estates. This was one of the earliest attempts at
community care for people who we would now
describe as having learning disabilities. Twenty-seven
people aged between 20 and 60 years old, who had
been discharged from a long-stay hospital, were placed
there to 'live in the community'. The hostel, with
bright yellow doors and luminous green window
frames, was surrounded by a 1.8-metre (6-foot)
wooden fence and had two enormous iron gates
guarding its entrance. Every member of staff had to
wear white coats while on duty, so all in all the hostel,
rather than providing care in the community, was
really just a smaller version of what had gone before,
in a supposedly secured area where the real estate was
considerably cheaper. Did I hear someone shout
'cynic'?

A warden and her deputy lived at the project in flats
attached to the hostel. Once again standing regula-
tions prohibited the keeping of pets but the warden
decided to get a dog anyway because this was her
home, and she wanted a canine companion. So she
duly acquired an 18-month-old golden retriever who
she called Oakie. It became apparent soon after his
arrival that he was no ordinary dog. Oakie was very
gentle with the residents and did all he could to put
them at their ease. However, one or two of them were
initially quite frightened and would sometimes run
away from him. He approached these people particularly carefully, doing his best not to cause them any
alarm. He would work on people individually, but

always using a similar tactic. Initially he would keep his distance but then gradually, day by day, he would move a little closer to them until he sat right by their side. His approach worked, even on the two most frightened residents, who soon became his greatest admirers and ended up being the ones who took most care of him. They used to brush him and give him biscuits that they had saved from their tea.

The hostel had a garden area but it was seldom used by the residents because of the abusive attentions of the neighbourhood children. If they became aware that residents had gone into the garden, they would climb the wooden fence, call them names and sometimes throw stones at them. On the rare occasions that residents did venture into the area they would invariably retreat very quickly to the house, terrified and often in tears. However, all this changed when Oakie arrived. At first the residents used to watch him walking around the garden and then, a little hesitantly at first, they began to follow him out. The local children, unaware of Oakie's presence continued to climb the fence, but he had an uncanny ability to predict exactly where the assailant's head would appear and he would be up at them in a flash barking and growling. As one would expect, he had some success at deterring future assaults and the residents began to feel safe in the garden when he was with them.

Of course not all of the children on the estate were badly behaved and before too long they were coming to the iron gates looking for the dog. Oakie, without barking or growling, would cautiously approach them and stop by the gate. He appeared to be sizing them up and, when he seemed satisfied that they were no threat, he would let them stroke him. He also used to roll on to his back and let the children rub his belly. His behaviour with these children was witnessed by the residents and it gave them the confidence to approach the children too. Oakie would go to the children and then look back at the residents and it was almost as though he was saying 'Come with me'. Some of the residents would then join him at the gates and begin talking with the children. As time went by they began to call each other by name and friendships were made.

The whole atmosphere on the estate slowly began to change. Residents were greeted in the street by the children and nearly all of the abuse stopped. At weekends the staff began letting the children into the grounds to play football or just to have a drink. Parents also started to come over, initially just to see what their kids were up to, and Oakie would give them a very warm welcome. Eventually they too began regularly popping in at the weekends. It wasn't long before the staff and residents had an open day where they were joined by many of their neighbours. Attitudes had changed and so had the residents. They were more relaxed, happier and for many of them, for the first time in their lives, they felt part of the community.

The residents' new neighbourhood friends were very supportive and they themselves became very protective of them. I suppose in many ways they were just taking a leaf out of Oakie's book. Without doubt this remarkable dog brought this community together in a way in which no human would have been capable. Even though Oakie is no longer with us, his spirit lives on. The children who first knew him are now the adults of the estate with small children of their own. Their attitudes haven't changed and they continue to be very supportive of the current hostel residents. One can be fairly optimistic that these new parents will encourage their children to behave in a similar way so

Oakie's legacy will not be lost and, hopefully, these same attitudes will be embraced by many future generations to come.

PART III

WHAT NEXT?

CHAPTER 10

FUTURE DIRECTIONS

In the previous chapters I have tried to offer some insights into the world of animal assisted therapy and its development over the past several decades. I would now like to describe some of the possible directions it may follow as we progress into the next century. To complete the picture I would also like offer one or two personal reflections on past and current activities in this field.

It would certainly seem that the exploration into the potential of dolphin therapy has only just begun. As advances are made in the electronic collection of neurological data, we may gain greater insights into the therapeutic dynamics of human encounters with dolphins. The findings of initial studies have indicated that dolphin sonar may have a relaxing effect on humans but further research is necessary to acquire a fuller understanding. A very recent development in the field may do much to speed this process along. The Aqua Thought Foundation in the USA has produced MindSet™, a 16-channel neuromapping system for the collection of brain wave data before and after interaction with dolphins. The effectiveness and application of the system are currently being evaluated. This piece of equipment has the potential to identify both the neurological and psychological factors that may contribute to the therapeutic process.

Much of Dr David Nathanson's dolphin human therapy (DHT) has centred around working with children with disabilities to enhance their learning potential. However, in August 1998, he began working with an 11-year-old boy, Jo Harrison, who suffered from Tourette's syndrome. Typical symptoms of this incurable genetic disorder include facial tics and physical jerks and grunting, which often develop in time into compulsive outbursts of swearing. This was a first for Dr Nathanson and he had no idea if dolphin human therapy would be of any benefit. However, after 14 sessions with the dolphins over a two-week period, the average number of body tics the young boy experienced in an hour fell from 48 to three. This unexpected outcome opens up a new avenue of exploration which I'm sure Dr Nathanson will be pursuing with some vigour.

An innovative approach to holistic healing involving dolphin related therapy has recently been developed by Pat Morell. After many years of practising her own very gentle form of reflexology, she now offers a treatment described as aqua-reflex. This is a further refinement to her personal method which is conducted in water with dolphins present or using dolphin visualisations. Aqua-reflex is somewhat more than just practising reflexology in water. Pat certainly takes people's feet into her hands but she only administers the most gentle of treatments for maybe three minutes or so, concentrating on very specific areas. She firmly believes that her work impacts on the spiritual, intellectual and physical being. Her method utilises the benefits of a water borne treatment and the links between the natural energies of dolphins, the oceans and the human world. Pat's work is drawing an increasing amount of attention and she has reported significant success working with a range of conditions but especially in the treatment of partial physical paralysis and emotional difficulties.

A number of observers have expressed concerns about the use of dolphins for therapeutic purposes, both in the wild and in captivity, because of the possible detrimental effects on the creatures and their living environment. Others have highlighted the purely practical difficulties of large numbers of people attempting to make close contact with dolphins. These concerns are not without foundation and, as a result, some thought-provoking means of simulating dolphin encounters and experiences have been proposed. One of the earliest advocates of this approach was Dr Horace Dobbs. During the 1980s and 90s he proposed several ideas on the subject. His first project was a 'dolphin audio pill', which captured the sounds of a dolphin encounter to lift the spirits. This was developed into a commercially available tape and CD, *Dolphin Dreamtime*, which featured the sounds of whales, dolphins and the sea with a human voice-over to guide the listener. Everyone who listened to the tape was asked to describe their personal response for evaluation purposes. Reports indicated that listeners were certainly relaxed by the experience and it helped many achieve what was described as a 'state of reverie'.

In the early 1990s Dr Dobbs proposed the development of Dolphin Therapy Centres to help those suffering from mental health problems. The centres were to offer a range of counselling services and complementary therapies, but the focal point would be a dolphin therapy pool where people would be able to experience simulated dolphin encounters and experiences. The pools would be filled with isotonic salt water maintained at body temperature. Dolphin images would be projected on to the water and their sounds were to be transmitted through underwater loudspeakers. To date, as far as I could ascertain, no Dolphin Therapy Centres have been built or are under construction.

The most recent simulation project reported by Dr Dobbs involves the creation of Dolphin Domes. A prototype was constructed in 1996 by Ken Shapely. This took the form of a dome shaped polyhedral structure covered in a translucent material decorated with dolphin images. A didgeridoo was played inside the dome to create 'sound images of dolphins'.

To be honest, I find much of Dr Dobb's work, which I'm sure is well intentioned, a little too mystical for my tastes. I am far more attracted to the idea of information technology-based virtual reality simulations. Advances in this area and the development of the neurophone (a device for sending a sound signal which can be received by humans with no discernible external audio signal present) have enabled the creation of a sophisticated computer generated dolphin encounter simulator known as Cyberfin™. Only time will tell, however, if any of these simulations can offer the same benefits as those derived from the real thing.

A Chicago based animal assisted therapy group, known as Chenny Troupe, leads the world in developing a most valuable new area of support dog work. This group has used selected dogs in a range of standard therapeutic programmes for several years. Since 1993 however, they have rendered a unique service to their community through their involvement in a drug rehabilitation project known as City Girls. Chenny Troupe volunteers and their dogs offer an additional assistance to teenage girls on the programme. The dogs provide supportive and positive experiences for the girls as they broach problems in their human relationships and confront challenging issues related to trust, commitment, healthy relationships and risk. The girls are also given an opportunity to develop nurturing and control skills by grooming and walking the dogs. Many of the girls have had very little control over their lives and this experience proves invaluable for their future personal development.

Chenny Troupe's outstanding contribution to this programme was recognised when they received the 1998 Amoco Leader Award in health/substance abuse. The award was accompanied by a cheque for $50,000. The money is to be used to develop an animal assisted therapy model that can be used throughout the USA. This innovative scheme thoroughly deserved its award and it is hoped that their work can be emulated in Britain and throughout the world.

One of the most fascinating developments in the world of assistance dogs has been the identification of canine companions who have the ability to detect the onset of epileptic seizures. The positive applications of this capability are self-evident, but research is required to determine whether this is a skill that most dogs could acquire through training or whether this is an innate ability that cannot be taught. Whatever the answer, this is surely just the beginning of an enhanced human understanding of dogs' support and assistance potential. For example, recent anecdotal reports have described dogs that have repeatedly nudged their owners in certain areas of their body, which have later been identified as the site of a tumour. The future of this field appears most exciting and I have a feeling that my next foray into research may well be in this area.

As far as our feline friends are concerned, time will tell just how many of them make the grade as PAT cats. Even though our cat Bebe is as affectionate as many dogs, he wouldn't stand for being carted from place to place for visits. I'm no expert, but most cats, in my experience, find it difficult to adjust quickly to new environments. This may sound rather pessimistic but, if I'm proved wrong, I'll be very happy to publicly acknowledge the fact. For those cats that do possess the required qualities, the best of luck to you and congratulations on an excellent job.

This seems an appropriate point to acknowledge June McNicholas for her innovative approach to animal assisted therapy. Her introduction of ferrets into this field of endeavour should be commended. This much maligned species has often received negative publicity as a result of their use and abuse by humans. My only experience of these creatures has been restricted to witnessing one or two very strange men putting a ferret or ferrets into their pants or trousers. I may be getting old but this seems to have very limited entertainment value (that's unless, of course, one of the creatures is a tad peckish and decides to snack on a source of nutrition close to hand). However, I wish June well and I'm very much looking forward to hearing more about these little charmers and their therapeutic contribution.

Of all the programmes described within these pages, one in particular has left many observers with the most ambivalent of feelings. This concerns the use of capuchin monkeys. While fully recognising the tremendous practical and emotional support they offer to quadriplegic people, reservations have been expressed by animal welfare activists regarding the plight of the monkeys. Many have felt particularly uneasy about the filing and possible extraction of their teeth. It has been argued that if animals are deemed to require such interventions to be acceptable in human social and physical environments, one could question whether these surroundings are indeed appropriate for them.

The future of animal assisted therapy (AAT) may prove to be dependent on the development of accredited training programmes similar to those available for the more established care professions. Much of the work in this field is currently undertaken by volunteers who have had limited opportunities for training. Support and advice on the implementation of AAT programmes are obtainable from The Society of Companion Animal Studies in Britain and from the Delta Society in the USA, but little is available in the way of specific training. In August 1998 a professional course in AAT was established in Israel and it may prove to be the first of its type in the world. Sandra Stone from the Children in Hospital and Animal Therapy Association (CHATA) is keen to see the implementation of AAT training programmes to further develop her work. The appropriate training and close supervision of practitioners is crucial to maintaining the required standards of practice in all the caring professions. As we are all surely aware, this need is never greater than when professionals are working with children. Advocates of AAT would claim no exception. In fact, given the nature of the work, more vigorous standards may need to be applied.

Sister Chiara Hatton-Hall, the International Liaison Officer for the Riding for the Disabled Association, has highlighted the need for training developments in her field. She is keen to see riding therapy training become available in mainstream colleges. She would also like to see Hippotherapy (physiotherapy on horseback) receive a similar level of recognition in Britain as it does in the USA, Germany and France.

The spread of HIV and AIDS has proved to be one of the most significant threats this century to the health and well-being of many within our society. The support companion animals have offered to many of those affected by these conditions should not be underestimated. A client of mine, Kurt, who tested HIV positive in October 1994, found the company of his animals helped him begin to come to terms with his situation. It is with his full permission that I present some of his thoughts:

'In my room at that time when my life was dark, keeping animals gave me an interest with responsibility. Holding my pets in my arms allowed me to release pain stored deep inside. Their comforting spirituality somehow enabled me to concentrate on the positive aspects of my life. The walls slowly started to crumble and my existence began to feel worthy again.'

Kurt was keen to share his thoughts because he felt others in a similar situation to his own may be able to derive some comfort from companion animals too.

One of the most pleasing events to have taken place during the preparation period for writing this book was the news that a substantial sum of money had been allocated by the British Lottery Commission to enable The Society for Companion Studies (SCAS) to further its pet bereavement work. Their counselling service for people who have been recently bereaved through losing a pet has been in existence for some time. The Lottery money will enable an enhancement of the service in conjunction with the Blue Cross animal welfare organisation. This work is of great importance to the development of AAT. If one recognises the close bonds that can be established with companion animals, one must be equally aware of the pain suffered when such animals are lost or die. Feelings of bereavement are also suffered by a good number of elderly people who have no choice but to give up their pets when they move into care homes or sheltered environments. Thankfully moves are afoot to address this issue but it will be some time before an abundance of supported accommodation which welcomes both the elderly and their pets becomes widely available. In England, the Anchor Housing Trust is at the forefront of a drive to meet this need and they are to be commended for their efforts.

Many animal assisted therapy projects and initiatives have not been so lucky in their pursuit of funds. Money is not freely available for these programmes, though a number of groups have received limited sponsorship support from pet food manufacturers and other commercial enterprises who may benefit from such promotions. Financial stability and solvency are perennial problems for most groups. CHATA, for example, receives a small amount of sponsorship money and occasional donations but it manages to survive only because the founders foot many of the bills. There is never enough money to meet the needs of many worthy causes but it seems shameful that such a cost effective approach to enhancing health should be so under funded. Many animals used in programmes are from rescues and the costs of caring and feeding animals in comparison to other interventions are negligible. I am sure a thorough cost/benefit analysis would support my contentions and reveal a healthcare initiative that unquestionably deserves financial assistance.

Throughout the pages of this book I have described the differing ways in which animals can contribute to the well-being and health of the human population. So before the final curtain, I would like to close with a tale of how a group of people, many of whom were chronically ill themselves, rallied to the aid of an ailing animal.

In 1996 Biba, the then 14-year-old visiting PAT dog at The Royal Star and Garter Home, suffered the canine equivalent of a very severe stroke. Almost unable to move and desperately ill, she was rushed to a veterinary hospital. Her owner, Marion, and the staff and residents of the home were deeply concerned about her condition. Her life was in the balance but after five days of round-the-clock attention she began to pull through which, given her advancing years, was close to a miracle. She returned home in a very weakened state, still requiring close attention. After a few days at home Marion took Biba and her younger dog Sonny to The Royal Star and Garter Home for their regular PAT visit. She didn't want to let the home down and she thought a change of scenery might be good for Biba. The dog was still extremely weak and Marion had no intention of walking her around the wards. She did consider putting her in a wheelchair and doing the rounds that way, but the patient wasn't keen. In the end Biba was gently placed on a cushion in the Home's very grand entrance hall to rest while Marion and Sonny made their visits. Well, the word must have got round that Biba was back because after more than ten years of visiting the disabled residents on the wards, they decided to return the favour and she was treated to a continuous stream of visitors as she lay in her makeshift bed. One very elderly resident who rarely left his ward insisted he be taken to see her. The nursing staff duly obliged and he was wheeled to the hall where he gave Biba the biscuit he had been saving for her. It wasn't just the residents who came. Many of the staff called on her to wish her well, including matron and the chief executive of the Home. Thankfully Biba made an almost complete recovery. Marion is convinced that the staff and residents made a more than significant contribution to her recuperation because they made her aware of just how very much she was loved.

Biba, now over 16 years old, is in pretty good health, but she does require regular medication for arthritis. She is a little weak in the back legs due to the condition but she now attends for acupuncture sessions which appear to be proving very effective in preventing any further deterioration. In fact, immediately following her treatments she has very noticeable energy bursts. Biba is such a placid and co-operative patient that her acupuncturist has enlisted her support as a demonstration dog when instructing trainees. In all her roles, Biba has been the most wonderful advertisement for both animal assisted therapy and a holistic approach to health. If any Royal readers will forgive me, to many Biba is very much the Queen of the Royal Star and Garter Home and long may she remain so.

CHAPTER 11

SOME PRACTICAL GUIDELINES

Within the next few pages are some general suggestions and advice for anyone considering implementing an AAT programme in a care environment. This is followed by details of the most appropriate ways to support working guide dogs and their clients in public places, courtesy of Alex who we met in Chapter 5. The final offering in this section is a description of the recommended behaviour to adopt when meeting dolphins both in and out of the water.

The General Idea

Consultation Suggestions for AAT programmes come from a variety of sources – patients, clients, residents, family members, professionals, in fact anyone who is involved in the support and care of those considered in need. I mention this because these are exactly the same people who need to be consulted before a programme is implemented. Consultation is crucial for a successful scheme.

Medical and Veterinarian Supervision The health and welfare of both the humans and the animals involved are paramount. No programme or project should be undertaken without the involvement of medical, veterinarian and other relevant care professionals (a psychologist, for example). Medical and veterinarian supervision are crucial for a safe scheme.

I would strongly recommend that anyone considering an AAT programme seeks advice from organisations who are experienced this field of endeavour. The two groups with which I am most familiar are the Delta Society based in the USA and the Society for Companion Animal Studies (SCAS) located in Scotland. (Their addresses are given in the contact section.) The aim of the Delta Society is to promote the use of animals to help people improve their health, independence and quality of life. The aims of SCAS are: *i)* to advance the understanding of relationships between people and companion animals; *ii)* to disseminate information about human/ companion animal relationships; and *iii)* to promote the quality of life of people and pets by encouraging responsible attitudes.

Both of these organisations belong to the International Association of Human Animal Interaction Organisations (IAHAIO). At their Prague conference in September 1998, IAHAIO adopted the following guidelines on animal assisted activities and AAT:

1) Only domestic animals which have been trained using techniques of positive reinforcement (reward system) and which have been, and will continue to be, properly housed and cared for, should be involved.

2) Safeguards must be in place to prevent adverse effects on the animals involved.

3) The involvement of assistance/therapy animals should be potentially beneficial in each case.

4) Basic standards should be in place to ensure safety, risk management, physical and emotional security, health, basic trust and freedom of choice, personal space, appropriate allocation of programme resources, appropriate workload, clearly defined roles, confidentiality, communication systems and training provision for all persons involved.

These policy guidelines are primarily intended for groups who are involved in AAT projects who may wish to join IAHAIO. For my purposes, the following suggestions, drawn mainly from SCAS recommendations on the introduction of animals into nursing homes and other institutions, may be of more practical use.

Firstly, consider the environment which will accommodate any proposed programme. What space and facilities are available? Are the needs of people who do not welcome the company of an animal being addressed? What are the possible implications for your particular client group? Will the space available limit the size and type of animal used? What are the toileting requirements? What are the hygiene implications? What other animals live in this environment/area already? Animal welfare organisations may help with such environmental evaluations.

Once the environmental implications have been thoroughly considered, three questions need to be addressed:

1) Would a visiting or resident pet be more appropriate?
2) Which type of pet is preferred?
3) What are the potential benefits and practical problems?

A visiting programme more readily addresses many of the practical issues but some would argue that those being cared for are denied the potential benefits derived from the constant companionship of a resident animal. Visiting animals are, however, more easily kept away from people who have allergies or phobias. There are no long-term care implications and patients/residents can choose from a wide range of animals. Another major advantage of visiting animals is that experienced organisations such as PAT dogs and cats already exist that can provide the service. Their visiting animals' health and temperament have already been assessed and their volunteers have undergone some basic training. Again these groups are well versed in working with care professionals and are familiar with the health and hygiene considerations in nursing environments.

A resident pet is an on-going commitment and someone must be ultimately responsible for their care. However, for some residents, the feeding, caring and nurturing involved can do much for their self-esteem, as has been clearly demonstrated. The type of pet selected is a most important consideration. It will determine the level and nature of the responsibilities associated with keeping the animal and will similarly affect health and hygiene considerations.

Do all you can to ensure any selected animal has an appropriate temperament for living with vulnerable people. Staff must be happy to share some of the responsibility. Staff and/or residents must also possess the necessary level of knowledge to ensure an animal is properly cared for. Before a final selection is made consult a veterinarian who can advise on the needs of particular animals and any potential problem areas. Most vets will do all they can to support such projects. In all cases a trial period would be recommended. Once a programme is established regular reviews are also advisable as needs continuously change. It is also necessary for animals to be regularly checked by the vet and proper care records kept. It is important too that animals have more than adequate rest and recreation periods. Due consideration of an animal's nutrition requirements is also essential.

Health And Hygiene

To ensure that residents of nursing and care homes can enjoy the company of animals, close consideration must be given to all areas of health and hygiene. It is of crucial importance to consult with medical experts and veterinarians to minimise the likelihood of problems.

Zoonoses – diseases that are transmissible from animals to men and women – fall into three main categories:

1) Infections from bites and scratches.
2) Other infections, eg campylobacter, salmonella, ornithosis, streptococcal disease.
3) Parasites, eg ringworm, scabies, toxocariasis and toxoplasmosis.

The chances of us contracting diseases from other animals are minimal if a few simple measures are followed.

- All bites and scratches should be quickly and thoroughly washed to prevent infection. These should be rare occurrences if animals have been carefully selected for suitable temperaments.
- Clean up promptly after your pet.
- Wear gloves if possible and carefully dispose of faeces.

- Wash hands thoroughly after cleaning or grooming your pet.
- Dogs and cats should be regularly wormed. Ask your vet for advice.
- All animals should have regular check-ups with the vet.
- Staff employed in care homes run by the statutory authorities should consult the standing operational procedures for the keeping of animals.

IF YOU HAVE ANY DOUBTS CONSULT THE APPROPRIATE PROFESSIONAL.

Some Ways and Means for Supporting Working Guide Dogs and their Clients

1) Do not attempt to attract a dog's attention when he or she is working. This can be extremely dangerous, especially when the dog is being relied upon to guide his/her client across the road or on to trains and buses.
2) Please do not talk about a visually impaired person or their dog behind their backs in a public place. This can be both disrespectful and potentially distracting.
3) Please do not be afraid to offer help. It's insensitive to let someone stand for a long period on a bus or train journey because they are unaware that there is a seat available.
4) If you decide to offer some help or you want to speak to a person who is clearly visually impaired, let them know you are there by touching them lightly on the arm and then speak to them face on. If you do not touch the person they may be unaware that you are addressing them.
5) If you or a child with you wants to stroke or pet a

guide dog, ASK FIRST before you make any attempt to gain the dog's attention. In certain situations the dog's client will be more than happy for you to make a fuss of his guide, but in others it may not be possible. The key is to ask, because any approaches made without first seeking the person's consent are potentially dangerous and are very likely to be met with a very firm 'No!'

How to Behave When Meeting Dolphins

In a Boat

- Dolphins often like to take a 'jacuzzi' in the wash of a propeller. They can get very close to the blades, so please take the greatest care in these situations when manoeuvring your craft. After seeing Freddie the dolphin's lacerations, which were caused by a boat almost backing in to him, I cannot emphasise enough the need for careful handling.
- If possible, let dolphins make the first move and come to you.
- Don't be too intrusive. If dolphins are feeding, mating or with their young, don't get too close or disturb them with unnecessary noise.

When Swimming

- Make sure you are aware of the local conditions, particularly strong currents, etc, and take all necessary precautions for safe swimming.
- Non-swimmers and weak swimmers should only enter the water wearing a buoyancy aid and when there is adequate support available to help them if

they begin to experience difficulties.
- Children should be supervised at all times.
- If you are taking any medication, consult with your doctor about any possible problems that could arise through immersion in cold water or exposure to strong sunlight.
- If you are pregnant or think you might be, consult with your doctor before swimming. Some organised dolphin swim programmes do not allow pregnant women to take part because the dolphins very quickly detect the second heartbeat and become extremely curious and attentive. As a result, other swimmers have very little chance of making any close contact with them!
- Take off rings and any other items of jewellery that could possibly cut or graze a dolphin.
- Let dolphins make the first move. If they come to you, keep still in the water at first and let them check you out.
- Only attempt to touch a dolphin when you feel your attentions are being invited.
- Do not touch a dolphin's face, mouth, eyes, blowhole or the tiny pinprick ears behind the eyes. Gentle stroking of the creature's belly and flanks are usually acceptable. Do not make any sudden movements or grab at their fins.
- Male dolphins sometimes extend their penis when in the company of humans. This is quite usual dolphin social behaviour, so try not to be alarmed. They may even attempt to tow you with the appendage. If you feel uncomfortable with such an interaction slowly withdraw and return to land or boat.
- If a dolphin appears to be becoming aggressive by nipping, butting or ramming, keep your arms at your side and calmly swim away.

CHAPTER 12

CONTACT ADDRESSES

L isted in this section are contact address for various organisations mostly in the United States that are involved with Animal Assisted Therapy or the general well-being of animals. At the time of going to print, the information is correct as far as I am aware. However, contact details do change regularly so I apologise in advance for any that are now out of date.

Assistant Dogs

Canine Companions for Independence
PO Box 446
Santa Rosa, CA 95403

Chenny Troupe
1504 North Wells Street
Chicago, IL 60610

The Seeing Eye Inc.
PO Box 375
Morristown, NJ 07963–0375
Tel: (973) 539–4425
Fax: (973) 539–0922

Dolphin Therapy

Dolphins Plus / Island Dolphin Care
PO Box 2728
Key Largo, FL 33037
Tel: (305) 451–1993
e-mail: dolphins-plus@pennekamp.com
www: http://www.pennekamp.com/dolphins-plus
http://www.islanddolphincare.org

Dolphin Human Therapy
13615 South Dixie Highway #523
Miami, FL 33176–7252
Tel: (305) 378–8670

Human Dolphin Institute
8317 Front Beach Road
Unit 37 A2
Panama City Beach, FL 32408
Tel: (850) 230–6030
e-mail: hdolphin@beaches.net
www: http://www.human-dolphin.com

Dolphin Swimming (Including Travel Companies)

The Caribbean Dolphin Experience
522 Pixie Trail
Mill Valley, CA 94941
Tel: (415) 388–1772
e-mail: JacquelineSchwartz@compuserve.com

Dancing Dolphin Institute
PO Box 959
Kihei, HI 96753
Tel: (808) 879–7044
e-mail: angelic@maui.net

The Divine Dolphins
4703 Old San Jose Road
Soquel, CA 95073
Tel: (408) 464–1178
e-mail: cyberark@pacbell.net
www: http://www.cyberark.com/dolphin

Dolphin Connection
PO Box 2016
Kealakekua, HI 96765
Tel: (808) 328–7353
e-mail: dolphco@aloha.net

Dolphins and You (Lei Aloha Center)
PO Box 4277
WA Anae, HI 96792
Tel: (808) 696–4414
e-mail: dolphins4u@aol.com
www: http://www.pixi.com/~dolphins/

Dolphin Dream Trips
170 Lake Drive
8 Palm Beach Shores, FL 33404
Tel: (561) 881–9124

The Dolphin Institute
5901 40th Avenue SW
Suite 3
Seattle, WA 98136
Tel: (206) 938–1263

Dolphinswim
PO Box 8653
Santa Fe, NM 87504
Tel: (505) 534–4188
e-mail: seaswim@bellsouth.net

Dolphin Watch
PO Box 4821
Key West, FL 33041
Tel: (305) 294–6306

Ecosummer Expeditions
1516 Duranleau Street
Vancouver, British Columbia
Canada V6H 3S4
Tel: (604) 669–7741
e-mail: trips@ecosummer.com

Isle of Shoals Steamship Co.
315 Market Street
PO Box 311
Portsmouth, NH 03802–0311
Tel: (603) 431–5500
e-mail: isleofshoals@rscs.net

Natural Habitat Adventures
2945 Center Green Court
Suite H
Boulder, CO 80301–9539
Tel: (800) 543–8917

Oceanic Society Expeditions
Fort Mason Center
Building E
San Francisco, CA 94123
Tel: (415) 441–1106

Sea Quest Expeditions
Zaetic Research
PO Box 2424
Friday Harbor, WA 98250
Tel: (360) 378–5767
e-mail: seaquest@pacificrim.net

Shearwater Journeys
PO Box 190
Hollister, CA 95024
Tel: (408) 637–8527

Wild Quest
1075 Duval Street
212 Key West, FL 33040
Tel: (305) 294–0365
e-mail: wildquest@sprynet.com
www: http://www.wildquest.com

General Animal Welfare

The American Donkey and Mule Society Inc.
2901 North Elm Street
Denton, TX 76201

American Humane Association
9725 East Hampden
Denver, CO 80231

Canadian Donkey and Mule Association
Cedar Sands Farm
RR10
Brampton, Ontario
Canada L6V 3N2

The Donkey Sanctuary of Canada
RR#6
Guelph, Ontario
Canada N1H 6J3

RELATA (Red Latin Americana de Tracción Animal)
c/o FOMENTA
Ap Postal 95 Telcor
Sucursal Douglas Mejia
Managua, Nicaragua

Human Animal Interaction

California Veterinary Medical Association
1024 Country Club Drive
Moraga, CA 94556–1900
or PO Box 6000
San Francisco, CA 94160

Delta Society
289 Perimeter Road East
Renton, WA 98055–1329

Green Chimneys Children's Services Inc.
Putnam Lake Road
Caller Box 719
Brewster, NY 10509–0719
Tel: (914) 279–2995
e-mail: samross@gchimney.org
www: http://www.gchimney.org

The Latham Foundation
Latham Plaza Building
Clement and Schiller Streets
Alameda, CA 94501

Pets Are Wonderful Council (PAW)
500 North Michigan Avenue
Suite 200
Chicago, IL 60611

Riding Therapy

North American Riding for the Handicapped
PO Box 100
Ashburn, VA 22011
Tel: (800) 369–RIDE

This organisation can supply you with details of a therapeutic riding center in your area.

BIBLIOGRAPHY

ALLEN, L.D., & BUDSON, R.D., 'The clinical significance of pets in a psychiatric community residence'.
The American Journal of Social Psychiatry, Vol. 2 (4),
41–43, (1982)

ALTMAN, I & ROGOFF, B., 'World Views in Psychology: Trait, Interactional, Organismic and Trans-
actional Perspectives.' *Handbook of Environmental
Psychology*, Vol. 1 (New York: 1987)

ANDERSON, W., 'Pet Ownership & Risk Factors for
Cardiovascular Disease'. *Medical Journal of Australia*
(1992)

ANDRYSCO, R.M., *'Pet-facilitated therapy in a retirement
nursing care community'*. Paper presented at the first
International Conference on Human/Companion
Animal Bond, Philadelphia (unpublished) (1981)

ANDRYSCO, R.M., 'A study of ethologic and therapeutic factors of pet-facilitated therapy in a retire-
ment-nursing community'. Unpublished doctoral
dissertation, Ohio State University (1982)

ARKOW, P. (ED), *Dynamic relationships in practice: Animals
in the helping professions* (California: The Latham
Foundation, 1984)

ATKINSON, D., 'Nothing More Precious'. *Social Work
Today*, 16 (46), 13–14 (1985)

BAUN, M.M., BERGSTROM, N., LANGSTON, N.F. &
THOMAS, L., 'Physiological effects of petting dogs:
Influences of attachment'. In R. K. Anderson *et al*
(eds), *The Pet Connection* (University of Minnesota
Press: CENSHARE, 1984)

BECK, A.M., SERAYDARIAN, L. & HUNTER, G.F.,
'Use of animals in the rehabilitation of psychiatric
inpatients'. *Psychological Reports*, 38, 63–66 (1986)

BIEBER, N., 'The integration of a therapeutic equestrian program in the academic environment of children with physical and multiple handicaps'. In A.H. Katcher & A.M. Beck (eds) *New perspectives on our lives with companion animals* (Philadelphia: University of Pennsylvania Press, 1983)

BLAU, P., *Exchange and Power in Social Life* (New York: John Wiley & Sons, 1964)

BOWLBY, J., *Attachment* (Harmondsworth: Penguin Books, 1969)

BREAKWELL, G.M., *Interviewing* (Leicester: BPS Books & Routledge, 1990)

BRICKEL, C.M., 'The therapeutic roles of cat mascots with a hospital-based geriatric population'. A staff survey, *The Gerontologist*, 19. 368–372, (1979)

BRICKEL, C.M., 'Pet-facilitated psychotherapy: A theoretical explanation via attention shifts'. *Psychological Reports*, 50, 71–74 (1982)

BRICKEL, C.M., 'Depression in the Nursing home: A pilot study using pet-facilitated psychotherapy'. In R.K. Anderson *et al* (eds), *The Pet Connection* (University of Minnesota Press: CENSHARE, 1984)

BRICKEL, C.M., 'Pet-facilitated Therapies: A review of the Literature and Clinical Implementation Considerations'. *Clinical Gerontologist*, Vol. 15, 309–332 (1986)

BRIM, D.G. Jr, 'Socialization through the Life Cycle'. D.G. Brim & S. Wheeler (eds), *Socialization After Childhood* (New York: John Wiley & Sons, 1966)

CANTER, D. & CANTER, S. (EDS), *Designing for Therapeutic Environments: A review of research.* University of Surrey (New York: John Wiley & Sons, 1979)

COCHRANE, A. & CALLEN, K., *Dolphins and their Power to Heal* (London: Bloomsbury, 1992)

COLLIS, G.M. & McNICHOLAS, J., 'Health benefits of pet ownership: Attachment versus psychological support'. Paper presented at the 7th International Conference on Human-Animal Interactions: Animals, Health & the Quality of Life, Geneva (1995)

CONNELL, C.M. & LAGO, D., 'Favorable attitudes towards pets and happiness among the elderly'. In R.K. Anderson *et al* (eds), *The Pet Connection* (University of Minnesota Press: CENSHARE, 1984)

CORSON, S.A., CORSON, E. & GWYNNE, P.H., 'Pet-facilitated psychotherapy'. In R.S. Anderson (ed), *Pet Animals and Society* (London: Bailliere Tindall, 1975)

CORSON, S.A., CORSON, E., GWYNNE, P.H. & ARNOLD, L.E., 'Pet-facilitated psychotherapy in a hospital setting'. *Current Psychiatric Therapies*, 15, 277–286 (1975)

CORSON, S.A., CORSON, E., GWYNNE, P.H. & ARNOLD, L.E., 'Pet dogs as nonverbal communication links in hospital psychiatry'. *Comprehensive Psychiatry*, 18, 61–72 (1977)

CORSON, S.A., CORSON, E., 'Pets as mediators of therapy in custodial institutions and the aged'. In J.H. Masserman (ed), *Current Psychiatric Therapies*, Vol. 18 (New York: Grune & Stratton, 1978)

CUMMING, J. & CUMMING, C., *Ego & Milieu – Theory & Practice of Environmental Therapy* (New York: Adline Publishing, 1962)

CUMMINGS, S.R., PHILLIPS, S.L., WHEAT, M.E., BLACK, D., GOOSBY, E., WLODARCZK, D., TRAFTON, P., JERGENSEN, H., WINOGRAD, C.H. & HULLEY, S.B., 'Recovery of function after hip fracture: the role of social supports'. *Journal of the American Geriatric Society*, 36, 801–806 (1988)

CUSACK, O., *Pets and Mental Health* (New York: The Haworth Press, 1988)

DAVISON, G.C. & NEALE, J.M., *Abnormal Psychology 3rd Edition* (New York: John Wiley & Sons, 1982)

DePAUW, K.P., 'Therapeutic horseback riding in Europe and America'. In R. K. Anderson *et al* (eds), *The Pet Connection* (University of Minnesota Press: CENSHARE, 1984)

DEPUTTE, B.L. & BUSNEL, M., An Example of a Monkey Assistance Program: P.A.S.T. – The French Project of Simian Help to Quadriplegics. *Anthrozoos* 10 (2/3) (1997)

DISMUKE, R.P., 'Rehabilitative horseback riding for children with language disorders'. In R.K. Anderson *et al* (eds), *The Pet Connection* (University of Minnesota Press: CENSHARE, 1984)

DOBBS, H., *Tale of Two Dolphins* (London: Jonathan Cape, 1990)

DOBBS, H., *Dolphin Therapy Centres – A Vision for the Future* (Humberside: International Dolphin Watch, 1991)

DOBBS, H., *Dance to a Dolphin's Song* (London: Jonathan Cape, 1992)

DOBBS, H., *Journey into Dolphin Dreamtime* (London: Jonathan Cape, 1992)

DOBBS, H., *Dolphin Domes* (Humberside: International Dolphin Watch, 1997)

DOYLE, M.C., 'Rabbit – Therapeutic prescription'. *Perspectives in Psychiatric Care*, 13, 79–82 (1975)

ENDERS-SLEGERS, M-J., 'Do companion animals enhance quality of life for elderly people?' Dept. of Clinical & Health Psychology, University of Utrecht. Paper presented at the BPS Annual Conference, University of Warwick (1995)

FRANCIS, G., TURNER, J.T. & JOHNSON, S.B., 'Domestic animal visitation as therapy with adult home residents'. Unpublished paper, 1982

FRIEDMANN, E., THOMAS, S., NOCTOR, M. & KATCHER, A.H., 'Pet ownership and coronary heart disease patient survival'. *Circulation*, 58, 168 (supplement) (1978)

FRIEDMANN, E., KATCHER, A.H., LYNCH, J.J. & THOMAS, S.A., *'Animal companions and one year survival of patients after discharge from a coronary care unit'*. Public Health Reports, 95, 307–312 (1980)

FRIEDMANN, E., KATCHER, A.H., THOMAS, S., LYNCH, J.J. & MESSENT, P., 'Social interaction and blood pressure: Influence of animal companions'. *Journal of Nervous and Mental Disease*, 171, 461–465 (1983)

GEORGE, L.K., BLAZER, D.G., HUGHES, D.C. & FOWLER, N., 'Social support and the outcome of major depression'. *British Journal of Psychology*, 154, 478–485 (1989)

GLASS, T.A., MATCHAR, D.B., BELYEA, M., FEUSS-NER, J.R., 'Impact of social support on outcome in first stroke'. *Stroke*, 24, 64–70 (1993)

GRAHAM, B.M., 'Dolphin Therapy: An inquiry into the effects of close dolphin contact on clinically diagnosed sufferers of mental health problems'. Post-graduate research paper, Thames Valley University (1992)

GRAHAM, B.M., 'Enhancing the Therapeutic Environment: The Role of the 'PAT' (Pets as Therapy) Dog'. Master's Degree Dissertation, University of Surrey (1995)

GUNZBURG, H.C., & GUNZBURG, A.L., 'Normal environment with a plus for the mentally retarded'. In D. Canter & S. Canter (eds), *Designing for Therapeutic Environments: A review of research* (New York: John Wiley & Sons, 1979)

HAMA, H., YOGO, M. & MATSUYAMA, Y., 'Effects of Stroking Horses on Both Humans' and Horses' Heart-Rate Responses'. *Japanese Psychological Research*, 38 (2): 66–73 (1996)

HARLOW, H.F. & HARLOW, M.R., 'The Affectional Systems.' In A.M. Schrier, H.F.Harlow & F. Stollnitz (eds), *Behaviour of non-human primates*, Vol. 2 (New York: Academic Press, 1965)

HAUGHIE, E., MILNE, D. & ELLIOTT, V., 'An evaluation of companion pets with elderly psychiatric patients'. *Behavioural Psychology*, Vol. 20 (4), 367–372 (1992)

HENDY, H.M., 'Effects of pets on the sociability and health activities of nursing home residents'. In R. K. Anderson *et al* (eds), *The Pet Connection* (University of Minnesota Press: CENSHARE, 1984)

HOLCOMB, R. & MEACHAM, M. 'Effectiveness of animal assisted therapy program in an inpatient psychiatric unit'. *Anthrozoos*, Vol. 11, No 4, 259–264 (1989)

JENDRO, C., WATSON, C. & QUIGLEY, J., 'The effects of pets on the chronically ill elderly'. In R. K. Anderson *et al* (eds), *The Pet Connection* (University of Minnesota Press: CENSHARE, 1984)

JONES, M., '*Social Psychiatry in the Community, in Hospitals and in Prisons* (Springfield IU: Chas C. Thomas, 1962)

KAPLAN, S. & FREY TALBOT, J., 'Psychological Benefits of a Wilderness Experience'. *Human Behaviour and the Natural Environment*, Vol. 6 Chap. 5 pp. 163–201, Altman & Wohlwill (eds) (New York: Plenum Press, 1984)

KATCHER, A.H., SEGAL, H. & BECK, A.M., 'Contemplation of an aquarium for the reduction of anxiety'. In R.K. Anderson *et al* (eds), *The Pet Connection* (University of Minnesota Press: CENSHARE, 1984)

KELLERT, S., 'Affective, Cognitive, and Evaluative Perceptions of Animals'. *Human Behaviour and Environment: Advances in Theory & Research*, Vol. 6, 241–265 (1983)

KIDD, A.H. & FELDMAN, B.M., 'Pet ownership and self-perceptions of older people'. *Psychological Reports*, 48, 867–875 (1981)

KNOPF, R. 'Human Behaviour, Cognition and Affect in the Natural Environment'. *Handbook of Environmental Psychology*, Vol. 12 (New York: 1987)

LAGO, D., CONNELL, C.M. & KNIGHT, B., 'PACT (People and Animals Coming Together): A companion animal program'. In M.A. Smyer & M. Gatz (eds), *Mental Health & Aging: Programs and Evaluations* (Beverley Hills, CA: Sage Publications, 1983)

LANE, D.R., McNICHOLAS, J. & COLLIS, G.M., 'Dogs for the Disabled: Benefits to Recipients and Welfare of the Dog'. Paper presented to The World Veterinarian Congress (Japan: 1995)

LAWTON, M.P., MOSS, M. & MOLES, E. 'Pet Ownership: A research note'. *The Gerontologist*, 24, 208–210 (1984)

LEVINSON, B.M., 'The dog as co-therapist'. *Mental Hygiene*, 46, 59–65 (1962)

LEVINSON, B.M., 'Green chimneys seminar of plants, pets, people presents fresh perspectives'. *The Latham Letter*, p. 15 (1983)

LEVINSON, B.M., 'Human/Companion Animal Therapy'. *Journal of Contemporary Psychology*, Vol. 14 (2), 131–144 (1984)

LITTLEWOOD, J., *Aspects of Grief: Bereavement in Adult Life* (London: Routledge, 1992)

McNICHOLAS, J., COLLIS, G.M. & MORLEY, I.E., 'Social communication through a companion animal: the dog as a social catalyst'. Summary papers from proceedings of the International Congress for Applied Ethology (Berlin: 1993)

McNICHOLAS, J., COLLIS, G.M. & MORLEY, I.E., 'Pets and Health – supportive relationships?' BPS Annual Conference: Theoretical and practical implications of person-pet relationships (1995)

McNICHOLAS, J., 'P.A.T. Ferrets – Working Ferrets with a Difference'. *N.F.W.S. News* No. 40 (1997)

MUGFORD, R.A., & M'COMISKY, J., 'Some recent work on the value of caged birds with old people'. In R.S. Anderson (ed), *Pet animals and human development* (London: Bailliere Tindall, 1975)

NATHANSON, D.E. & DE FARIA, S., 'Cognitive Improvement of Children in Water and Without Dolphins'. *Anthrozoos*, Vol. 6, No 1 (1993)

NATHANSON, D.E., DE CASTRO, D., FRIEND, H. & McMAHON, M., 'Effectiveness of Short-Term

Dolphin-Assisted Therapy for Children with Severe Disabilities'. *Anthrozoos*, 10 (2/3) (1997)

NETTING, F.E., WILSON, C.C. & NEW, J.C., 'The Human-Animal Bond: Implications for Practice'. *Social Work*, Vol. 32 (1), 60–64 (1987)

NEWMAN, B.M. & NEWMAN, P., *Development through life: A psychological approach* (Homewood: Dorsey Press, 1984)

NORTH, F., SYME, S., FEENEY, A., SHIPLEY, M. *et al*, 'The Whitehall II Study: Psychological Work Environment and Sickness Absence Among British Civil Servants'. *American Journal of Public Health* Vol. 86 (3) 332–340 (1991)

OZCELIK, A. 'The Little Fish of Kangal: Licensed to Heal'. *Images of Turkey*, Issue 19 (1988)

ORY, M.G. & GOLDBERG, E.L., 'Pet possession and life satisfaction in elderly women'. In A.H. Katcher and A.M. Beck (eds), *New perspectives on our lives with companion animals* (Philadelphia: University of Pennsylvania Press, 1983)

PARSONS, T. & BALES, R.S. (EDS), *Family Socialization and Interaction Process* (New York: Free Press, 1955)

PEARL, R. *Dolphin Dreamtime Evaluation* (University of Swansea, 1994)

PERRIS, C., 'Integrating Psychotherapeutic Strategies in the treatment of young severely disturbed patients'. *Journal of Cognitive Psychology*, Vol. 6 (3), 205–219 (1992)

PILISUK, M., & PARKS, S.H., *The Healing Web: Social Networks and Human Survival* (Hanover, NH: University Press of New England, 1986)

REBER, A.S., *Dictionary of Psychology* (London: Penguin Books, 1985)

RIVLIN, L. & WOLFE, M., 'Understanding and evaluating therapeutic environments for children'. In D. Canter & S. Canter (eds), *Designing for Therapeutic Environments: A review of research* (New York: John Wiley & Sons, 1979)

ROBB, S.S., BOYD, M. & PRITASH, C.L., 'A wine bottle, plant, and puppy: Catalysts for social behaviour'. *Journal of Gerontological Nursing*, 6, 721–728 (1980)

ROBB, S.S., 'Health status correlates of pet-human association in a health-impaired population'. In A.H. Katcher and A.M. Beck (eds), *New perspectives on our lives with companion animals* (Philadelphia: University of Pennsylvania Press, 1983)

ROBB, S.S. & STEGMAN, C.E., 'Companion animals and elderly people: A challenge for evaluators of social support'. *The Gerontologist*, 23, 277–282 (1983)

ROSENTHAL, S.R., 'Risk exercise and the physically handicapped'. *Rehabilitative Literature*, 36, 144–149 (1975)

RUBERMAN, W., WEINBLATT, E., GOLDBERG, G.J., & CHAUHARY, B.S., 'Psycho-social influences on mortality after myocardial infarction'. *New England Journal of Medicine*, 311, 552–559 (1984)

SAVISHINSKY, J.S., 'Pet ideas: the domestication of animals, human behaviour, and human emotions'. In A.H. Katcher & A.M. Beck (eds), *New perspectives on our lives with companion animals* (Philadelphia: University of Pennsylvania Press, 1983)

SERPELL, J.A., 'Pet Psychotherapy'. *People-Animals-Environments*, Spring 1983, pp. 7–8 (1983)

SERPELL, J.A., *In the Company of Animals – A Study of Human Animal Relationships* (Cambridge University Press, 1986)

SERPELL, J.A., 'Beneficial effects of pet ownership on some aspects of human health and behaviour'. *Journal of the Royal Society of Medicine*, 84, 717–720 (1991)

SHELDRAKE, R. & SMART, P., 'Pets that know when their owners are coming home'. *Proceedings of Further Issues in Research in Companion Animal Studies Workshop*, at the University of Cambridge (1996)

SHEPARD, P. Jr, *Thinking Animals* (New York: Viking Press, 1978)

SMITH, B.A., 'Project inreach: A program to explore the ability of Atlantic Bottlenosed dolphins to elicit communication responses from autistic children'. In A.H. Katcher and A.M. Beck (eds), *New Perspectives on our lives with companion animals*, (Philadelphia: University of Pennsylvania Press, 1983)

SOCIETY FOR COMPANION ANIMAL STUDIES, *Guidelines for the Introduction of Pets in Nursing Homes and Other Institutions* (Glasgow: Straight line Publishing Ltd, 1991)

SOKOLOW, J., 'The Role of Animals in Children's Literature'. Unpublished manuscript, School of Forestry & Environmental Studies, Yale University (1980)

SOMMER, R. 'Dreams, Reality and the Future of Environmental Psychology'. *Handbook of Environmental Psychology*, Vol. 2 (New York: 1987)

STRAEDE, C.M. & GATES, G.R., 'Psychological Health in a Population of Australian Cat Owners'. *Anthrozoos* Vol. 6 No. 1, 30–42 (1993)

TIMM, C., 'Fourth International Congress for Therapeutic Riding'. *People-Animals-Environment*, Spring 1983, pp. 21–22 (1983)

TROUPE TALK, *This City Girl Gives Her All to City Girls* (Chicago: Chenny Troupe, 1998)

ULRICH, R.S., 'View Through a Window May Influence Recovery from Surgery'. *Science* 224, pp. 420–421 (1984)

VEEVERS, J.E., 'The social meaning of pets: Alternate roles for companion animals'. In M.B. Sussman (ed), *Pets and the Family*, pp. 11–30 (New York: Haworth Press, 1985)

WEBB, K., 'Four Good Legs'. Fourth International Congress on Therapeutic Riding (1982)

WEINBERGER, W., TIERNEY, W.M., BOOHER, P. & HINER, S.L., 'Social support, stress, and functional status in patients with osteoarthritis'. *Social Science & Medicine*, 30, 503–508 (1990)

WELLS, E., ROSEN, L. & WALSHAW, S., 'Use of Feral Cats in Psychotherapy'. *Anthrozoos*, Vol. 10 No. 2/3 (1997)

WORTMAN, C.B., 'Social support and the cancer patient: Conceptual and Methodological issues'. *Cancer*, 53, 2339–2360 (1984)

WHYAM, M., 'The Human animal Bond – Companion Animals in Prisons'. *The Society of Companion Animal Studies* (1997)

YATES, J., 'Project Pup: The perceived benefits to nursing home residents'. *Anthrozoos*, Vol. 1, No 3, 188–192 (1987)

YIN, R.K., *Case Study Research: Design and Methods 2nd Edition* (London: Sage Publications, 1994)

ZEE, A., 'Guide dogs and their owners: Assistance and friendship'. Paper presented at the International Conference on the Human-Animal Bond (Philadelphia, 1981)

INDEX